PENNSY ELECTRIC YEARS

Copyright © 1991
Morning Sun Books, Inc.

Published by
Morning Sun Books, Inc.
11 Sussex Court
Edison, N.J. 08820
Library of Congress Catalog Card Number: 90-063156
Cover design, layout, typesetting by R. J. Yanosey, Morning Sun Books.

First Printing
ISBN 1-878887-01-7

I would like to dedicate this book to the memory of my Dad, Theodore F. Volkmer who, after sending me to college, I'm sure would have preferred that I go into aerospace or nuclear physics rather than take up railroading as a profession. In later years he appeared gratified that I'd "made a go of it" in a field that I enjoyed and was glad that he did not discourage me from going into the profession I loved best.

Acknowledgements

There are many serious students of railroad history who spend countless hours combing through railroad archives and old newspapers, annual reports and the like dredging up facts, exploding myths and in general putting railroad history in its proper perspective. To those who have previously extensively researched the Pennsy electrification history, I owe a debt of gratitude for helping to make this book both interesting and more factual. One book in particular that I found to be very helpful was *Electric Traction on the Pennsylvania Railroad, 1895-1968* by Mike Bezilla published by Penn State Press. However the real credit for preserving a lot more history than facts and letters put to paper can ever attempt to preserve goes to that formerly small, but still growing number of dedicated railfans who have seen fit to expend the time and money to commit the railroad to film. Equally important to making an interesting and correct exposure is the photographer's dedication to recording the date and location of each photo subject and, where known, the train numbers. By doing this, researchers for years to come will be able to accurately determine when or why many obscure events took place merely by referencing properly located and dated photos. My sincere thanks go to those who not only exposed the film, but elected to share those films with the readers of this book. They include: Nelson Bowers, Rich Short, Al Roberts, Al Holtz, Ken Douglas, Bob Malinoski, Al Keller, Bill Brennan, Tony Schill, Fred Cheney, Bob Watson, Dave Sweetland, and of course, Bob Yanosey.

810516 $42.75
810516
PENN ELECTRIC YRS

Pennsy Electric Years

Electrification

During the early part of the 20th century, the Pennsylvania was by no means the only railroad to be interested in electrifying its tracks, but it was unquestionably the largest in terms of track miles being considered for electrification. The New Haven, Long Island, New York Central, the Boston & Maine (controlled by NH), the Norfolk & Western, and the Virginian were all experimenting with an alternative to the steam locomotive. Indeed, even the Erie and the Lackawanna had thoughts of electrifying.

There were two main objectives which fostered their quest for a replacement for the steam locomotive; elimination of large numbers of helpers on mountain grades, and elimination of smoke from the large cities. Pennsy found itself one of the few railroads in the nation beset with both problems. It should be remembered that in the 1900's, the 2-8-0 was king and it took a great many 2-8-0's to move freight over the mountains.

Today, in the 1990's as we look back at the Pennsy and other eastern railroads which are still electrified, we tend to think of the Pennsy as "that road that had all the locomotives," and we tend to look at the Lackawanna, the Long Island, and the Reading as those roads that only ran MU cars. In reality, the Pennsy was an MU ONLY property for over 15 years (1915 to 1930) before they really got seriously involved in A.C. overhead catenary powered long distance locomotive-hauled electrification. The New Haven, the B&O, the Virginian, the Norfolk and Western, the New York Central, the Milwaukee Road, Great Northern and the Boston and Maine were all veterans in the trade long before PRR came up with a design which seemed to work. To say that the Pennsylvania was conservative is certainly an understatement!

In parallel with the development of electric locomotives large groups of the railroad's engineers were engaged in perfecting the state-of-the-art of the steam locomotive. These were the same years that produced the Pennsy K-4, the I-1, the M-1 and the concept of superheated steam locomotion. As such, there was a great deal of competition for R&D dollars at the Philadelphia headquarters between the steam lobby and the pro-electrification forces.

A thumbnail chronological sketch of the history of the Pennsy electrification follows. Much more detailed histories, adequately illustrated with numerous black and white photographs have been previously published in a number of other fine books. Therefore, this book shall make no attempt to be the pre-eminent Pennsy electrification history. Rather, we shall in these pages describe in some degree of detail via color photos the state of the PRR electrification in the last ten or so years of the Pennsy (1958-1968) which happens to coincide exactly with the author's tenure as an employee of the railroad's Mechanical Department. It was this period when steam was phased out altogether and diesels brought in on the non-electrified portions of the railroad.

To Dieselize or Not To Dieselize

It might be safely said that it was the success of the diesel locomotive that ended once and for all the enormous rivalry between the steam and electric factions of the PRR and other railroad managements. The diesel, after all, contained the best of both worlds. It carried its fuel supply onboard wherever it went and was not dependent upon expensive power stations and labor-intensive catenary structures. At the same time it powered trains with the very same type of traction motors the electrical engineers so desperately desired on the electric locomotives.

A bonus with the diesel was that units could be multipled, thereby cutting crew costs. And, of course, the cities and tunnels instantly became smoke free, notably Pittsburgh where the sun hadn't been seen for generations. During the 1950's while PRR like other railroads was in the process of dieselizing, a whole lot of soul searching and debate was going on in Philadelphia to decide whether or not to scrap the electrification and dieselize everything or buy new electric locos. Were it not for the enormous investment in MU cars, diesels most certainly would have replaced the P-5's and GG-1's in the late 1950's.

The Locomotives

The most visible Pennsy electric locomotive in the '50's and '60's was unquestionably the GG-1. One hundred thirty-nine of them plied the rails with very, very little down time. But it was the non-GG-1 locos that gave the spice and flavor to the Pennsy electric scene. There were ninety P-5a boxcabs and "modifieds" which hauled the freight, largely at night when the passenger trains were fewer and farther between. These were phased out in the early 1960's and replaced with sixty-six E-44 type locomotives, each possessing roughly twice the pulling power of a P-5. Then there were the experimental freight electrics of the early 1950's era. There were six E-2b's built by G.E. which were capable of MU'ing with the P-5's. Also plying the rails were four Westinghouse rectifier locos, the two E-3b's and two E-2c's.

Lastly there were the plethora of relics left over from years gone by. There were two O-1's, three L-6's, two DD-1's, one DD-2, a P-5b, seven FF-2's and several surviving B-1's.

Given that much of the Pennsy electrified territory was "off limits" to railfans, not very many photos exist of the rarer breeds of electric locos. Another reason that Pennsy electric loco photos are relatively scarce is that many railfans consider the electrics as a sterile, lower form of life, so to speak. The electric locomotive didn't chuff, bellow clouds of smoke, make loud noises as they strained under load, or have melodious whistles. They just silently did their job, day in day out, rain, sleet, snow or sunshine. Incidentally, one of the first things new PRR employees were taught is there is nothing in the world that moves more silently and potentially more deadly than a GG-1! The same could be said about most any electric locomotive except the latter day rectifier units with their loud blowers.

On February 1, 1968, the day the curtain rang down on the mighty Pennsylvania Railroad, virtually all of the trackage ever electrified still remained intact. The only de-electrified trackage was the short Fort Washington Branch to White Marsh, Pa. Following is a chronology of "electrified" events which led to that famous day in Pennsy history.

Key Dates in PRR Electrification

****

1895 June 3. Burlington & Mt. Holly Branch of PRR is converted to D.C. trolley operation. First PRR electrification and first electrification of a steam railroad in the United States. Electric service is discontinued in 1901 when power house burns.

1901 December. PRR makes decision to go ahead with plans to construct tunnels under the Hudson and East Rivers to enter Manhattan and utilize electric locomotives. Increasing pressure from smoke abatement lobbies forces the decision.

1904 February. Construction begins on tunnels and Penn Station complex in New York City. Design for clearance overhead does not allow for overhead catenary installation.

1905 December. First two experimental electric locomotives, 10001 and 10002 are completed. Testing begins in early 1907 on the West Jersey and Seashore R.R. near Franklinville, N.J.

1908 July. Testing of overhead catenary configurations are conducted on 5 miles of Long Island R.R. using locomotive 10003 which was PRR's first straight A.C. loco. Pennsy in December 1908 decides to stay with D.C. third rail operation similar to LIRR and NYC as opposed to the AC operated catenary-powered New Haven and Boston & Maine railroads.

1909 April. First two-unit DD-1 is delivered from Altoona Works for testing.

1910 November 27. First electrically powered passenger train on the PRR departs Penn Station, New York.

1915 September 15. First Multiple Unit electric train operates from Philadelphia to Paoli. Represents first PRR A.C. electrification. Cost of electrification of 20 miles (93 track miles) is $4.2 million.

1917 April 1. Through passenger train service begins via Hell Gate Bridge. New Haven A.C. electrification ends at Sunnyside. Overhead catenary through Penn Station is still 13 years into the future!

1917 First PRR electric freight loco, the FF-1 *(Big Liz)* is outshopped. It carries road number 3931. At the time, Pennsy is considering electrifying freight trains over the Alleghenies. Success of I-1 and K-4 steam locos recently designed and tested dampens enthusiasm for continued expenditures for electrification. Loco is tested on the Paoli line which has a gradient.

1919 PRR President Samuel Rea decrees that no PRR Mechanical Officer shall rise to senior status without prior experience working in the electrified zone.

1924 Three L-5 2-8-2 locos are built. Number 3930 has A.C. propulsion and the other two, 3928 and 3929, have D.C. electrical gear. Railroad decides that a better financial return can be obtained by electrifying the New York to Philadelphia line rather than over the Alleghenies.

1926 Altoona constructs the BB-1, BB-2, and BB-3 switchers for yard service in New York and Philadelphia. Several more, the BB-2 class are built and D.C. -equipped for Long Island service. Fourteen more B-1's were built in 1934 bringing the total to 28 units. Sunnyside receives catenary. Altoona still building 3rd rail running L-5's as the wires go up!

1927 Altoona Works constructs 15 more L-5 class 2-8-2 D.C locomotives bringing the total to 23 such units. Many DD-1's are in turn sold off to the Long Island. The L-5's are less successful than the DD-1's they replaced. All were scrapped by 1942.

1928 Philadelphia to Wilmington is electrified in conjunction with the Broad Suburban Station modernization program.

1930 Hudson River tunnel receives catenary. Rails lowered for added clearance. Catenary is energized New York to New Brunswick and Philadelphia to Trenton. Pennsy decides it is time to get serious about developing long distance high speed A.C. passenger locomotives.

1930-31 Four pairs of O-1 class 4-4-4 electric locos are built. Each pair is designated a different sub-class, O-1, O-1a, O-1b, O-1c and each pair has minor variations in traction motor size. They are found to be too light for even the short passenger trains such as the LV trains to and from New York.

1931 June. Orders are placed for twenty five P-5a's. They become the first PRR electric locos to be built outside of Altoona. They are built at General Electric at Erie, Pa., and are placed into test service between New York and Wilmington.

1931 Fall. Pennsy contracts with Lima Locomotive Works for 30 L-6 class 2-8-2 locos. Two prototypes are built at Altoona. These are to be freight counterparts to the P-5's.

1932 Ninety P-5a's (in addition to the two P-5's, 7898-7899) are on order for eventual New York to Washington passenger service. 7898 is renumbered 4700 and 7899 becomes the 4791.

1932 December 9. MU car service begins between Manhattan and New Brunswick, N.J.

1933 January 16. First electric train runs New York to Philadelphia. P-5a 4764 hauls first northbound train from Philadelphia.

1933 April 10. All passenger trains to Harrisburg and the West are electric powered as far as Paoli.

1933 Serious problems arise with the poor riding qualities of the P-5's. A replacement design is sought with a high degree of urgency.

1933 Pennsy borrows a New Haven EP-3 loco (0354) to test on the mainline at Claymont, Delaware. PRR is impressed by the 2-C+C-2 wheel arrangement. Enter the GG-1 design.

1934 January 3. P-5a 4772, a boxcab loco strikes a truck at a grade crossing. All future P-5a's are built to a steeple cab configuration. 4770 is later rebuilt to this style but with a welded rather than a riveted carbody.

1934 August. Baldwin-Westinghouse produces a 4-8-4 electric loco number 4800. It is classed R-1. At the same time General Electric turns out a 4-6-6-4 loco number 4899 and it is classified GG-1.

1934 November. Railroad decides GG-1 superior to R-1 and orders 57 more GG-1's, 4801-4857. The 26 as yet unbuilt carbodies of the Lima L-6's are cancelled and the P-5a boxcab locos are assigned to freight duties in the New York Region. (GG-1 4899 is renumbered to 4800).

1935 February 10. Most New York-Washington trains become electrically powered using P-5's. By April 8th all trains are electrically powered and many use GG-1's. Arrival of additional GG-1's free up P-5's to pull freights to Washington.

1938 January 15. Electric passenger service is extended from Paoli to Harrisburg. By Spring the Low Grade and Columbia and Port Deposit lines are also electrified.

1938 Experimental freight version of the GG-1, called DD-2 (4-4-4-4) #5800 is built at Altoona as a prototype for Middle Division electrification then being proposed.

1942 PRR's contribution to the war effort results in the scrapping of 23 L-5's, some DD-1's and the 29 unfinished L-6a carbodies stored at Altoona.

1943 139th and last GG-1 is outshopped from Altoona. Because WWII is in progress, all consideration of electrification to Pittsburgh is shelved.

1948 Pennsy commits all available resources to the elimination of steam locos by substituting diesels as fast as money becomes available.

1949 PRR discards the idea of a "Super GG-1" and orders four experimental locos each from Westinghouse and General Electric.

1951 June. Locos 4939-4942, class E-2b, are delivered from G.E. as potential replacement for the P-5a's in freight service.

1951 November. Locos 4995-4998, class E-3b and E-2c are delivered from Westinghouse. These represent the first ignitron equipped electric locomotives with diesel type traction motors.

1954 Neither type of experimental locos are considered successful. The P-5a's soldier on, augmented by regeared GG-1's in freight service. Westinghouse exits the locomotive business. G.E. goes to work on an improved rectifier loco, delivers 10 to New Haven and 12 to the Virginian Railroad in 1955-56. PRR watches progress closely, is impressed by performance.

1956 Great Northern R.R. dieselizes its Cascade Tunnel operation. Pennsy, desperately short of freight power, buys seven 2-6-6-2 locos from GN, which it dubs FF-2's. Numbers them 1-7 and places them in helper service at Thorndale and Columbia. They are restricted to 35 m.p.h. because of low gearing.

1959 With last steam loco removed from service previously, PRR considers de-electrifying all non-MU operated lines. Decides against same and orders 66 E-44's from General Electric. Electrification to Pavonia Yard briefly de-energized.

1960 October 25. E-44 # 4400 is delivered and put on display at 30th Street Station, Philadelphia. Locos are delivered from Erie at the rate of two per month.

1962 August. After 36 E-44's are delivered, #4460 is delivered out of sequence as the first solid state rectifier loco. 4460-4465 are classed E-44a.

1963 Passenger Service Improvement Corporation (PSIC) buys *Silverliner* cars, leases to PRR and Reading.

1964 April. The last P-5a goes to scrap. Only one, the 4700, is preserved. It goes to the St. Louis Museum of Transport. All the 1951 experimental locos also go to scrap.

1965 Pennsy decides to upgrade passenger service from New York to Washington with high speed MU cars called *"Metroliners"* rather than buy replacement locos for the GG-1's. Orders 50 cars from the Budd Company. Coaches are Westinghouse-equipped, Parlors and Diners are G.E.-equipped.

1966 Four high-speed Budd test cars begin testing in New York Region.

1967 *Metroliner* deliveries begin from Budd Company. Some testing carried out on Reading Railroad. Electrical problems prove to be extensive and cars found to be too unreliable to enter revenue service.

1968 February 1. Pennsy flag is hauled down at #6 Penn Center Plaza and the Penn Central era is ushered in. The *Metroliners* had yet to see revenue service as of this date.

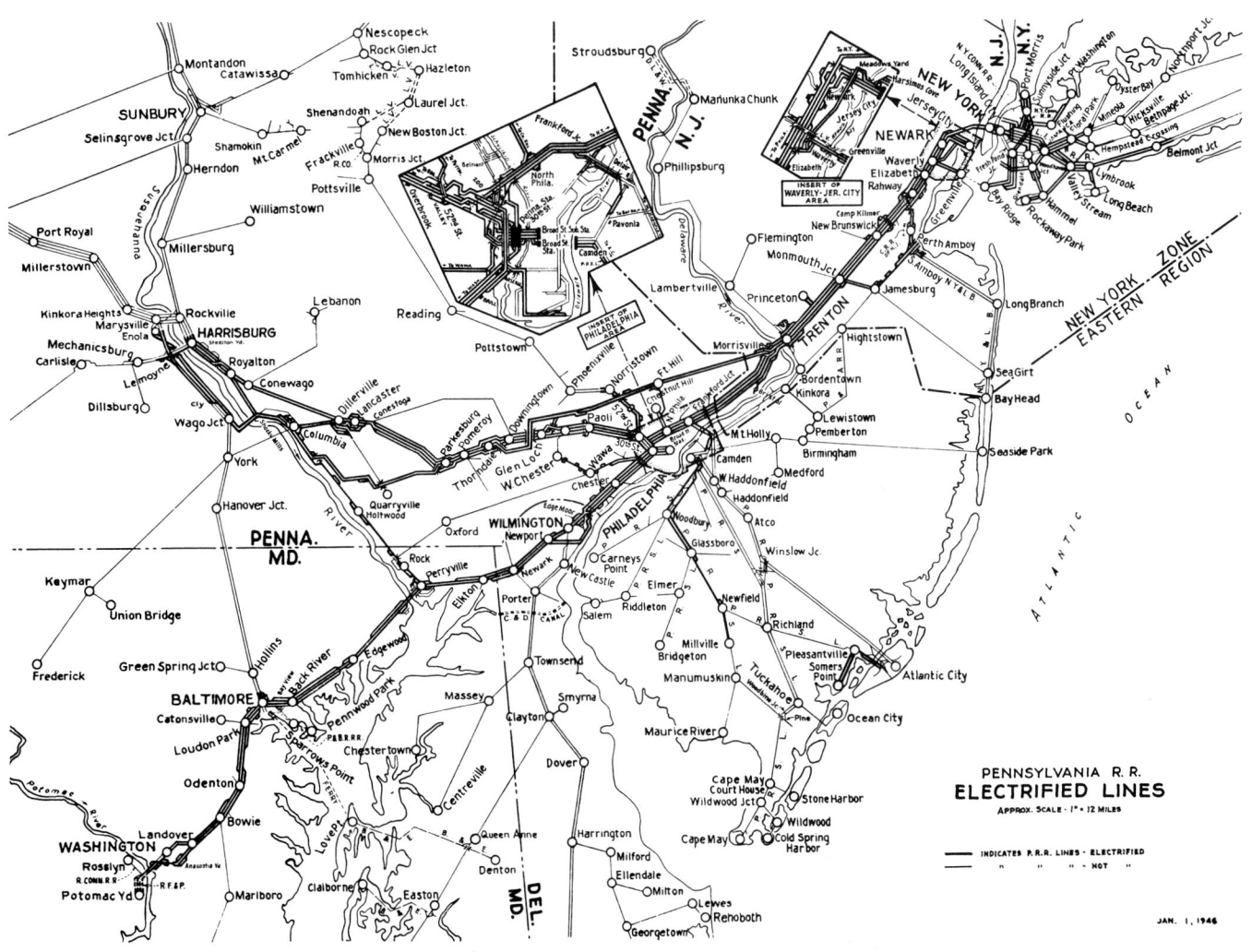

Standard Railroad of the World?

It has been pointed out in the pages of *Pennsy Diesel Years* that about the only thing "standard" about the Pennsylvania Railroad was its standard gauge track and that it operated on standard time most of the time between October and April.

The various idiosyncrosies of the dieselized PRR were merely an outgrowth of the evolution of the electric locomotive. The PRR seemed to always try out electric locomotives in groups of two or three, adopt one and throw away the other design, but not the loco. They seldom ever threw anything that resembled rolling stock away. Rather, they simply looked around for a use for it.

In the early years they would build an A.C. version and a D.C. version of a given locomotive. They decided D.C. was better, so they built 66 DD-1's, only to make the decision to electrify with A.C. Then in 1930 the PRR built three locomotive designs more or less in parallel. They were the O-1, the L-6, and the P-5 type. The idea was a noble one. Have the carbodies and motors interchangeable but just vary the wheel arrangement, weight on drivers, and gear ratio and *voila*, they'd have a light-duty passenger, a heavy-duty passenger, and a freight loco, all look-alikes. As we all know, it didn't work out that way. The O-1 was too light-weight for even short passenger trains (the MU's after all were the "short" passenger trains), the P-5 just didn't cut it for a variety of reasons on the heavy passenger trains, particularly coming out of New York City on the steep grade in the tunnel and the L-6's were surplussed when the GG-1 design bumped the P-5's into freight service.

The same theory held true for the GG-1 versus the R-1 runoff. The GG-1 was chosen as superior to the R-1 as the successful passenger loco of record in the 1934-35 time period. The 4-8-4 configuration R-1 was never duplicated because it couldn't negotiate tight curves in enginehouse territory, but they didn't scrap the 4999 either. Rather, they ran it for more than 20 years in passenger service, partially as a tribute to interchangabiltity of certain parts with the GG-1's.

In 1951, they did it again. They had two experimental freight locomotive designs built and tested. Again, one A.C. and the other with D.C. traction. Neither design was a resounding success. But did they throw them away? Of course not. The locos hobbled through eleven years of life as orphans of the fleet.

When I worked on the Pennsy, I used to quip that the only time the railroad would scrap a locomotive was when they discovered they had two alike! So much for standarization.

In all fairness to the Pennsylvania and railroads in general, there is a rational explanation for all this perceived madness. It is simply this; changing a large railroad overnight to take advantage of rapidly changing technology is a lot like trying to stop a huge mass while it is in motion.

The PRR could in no way convert large quantities of locomotives, freight cars etc. to one specific configuration no matter how technologically attractive it was. The conversion from steam to diesel was probably the most drastic surgery the Pennsy, or any railroad, ever did in its history. And that took ten years to complete. Even with ten years of conversion ending in 1957, ten years later (1968) those diesels were largely still being maintained in antiquated steam roundhouses across the system because there was not enough revenue being generated with which to build diesel oriented maintenance facilities.

On the *(opposite page)* the similarities of the O-1, L-6 and P-5 experiment are explored. At the *(top)* is L-6 #5939 at Penn Station, New York, March 2, 1960 with its pantograph almost collapsed under the low A.C. catenary at Penn Station. Catenary was an after-thought and was added only after the

tracks were dug out and lowered a foot for clearance. The L-6 differed from the O-1 and P-5 in that it had a single traction motor per axle, directly geared like a diesel as opposed to the O-1 and P-5's quill drive arrangement. Standardization?

(Center) This photo shows P-5a #4705 at Wilmington Enginehouse in a similar view for comparison, November 6, 1960.

(Bottom) O-1a 7853 is seen at the New York City Post Office between switching moves, March 2, 1960. *(All- Bill Volkmer)*

A few notes about Classification

The Pennsylvania Railroad classified their steam locomotives according to the Whyte system. Thus an 0-4-0 was an "A," an 0-6-0 was a "B," an 0-8-0 a "C," a 4-4-0 was a "D" and so on. Electric locos on the other hand were classified by other railroads by their motored wheel arrangement. For example, a GG-1 was considered in the industry as a 2-C+C-2. The + sign indicated an articulation pin, the "C" meaning three powered axles and the 2 meaning two unpowered axles.

The PRR, on the other hand, chose to classify its electric locomotives by their steam locomotive counterpart wheel arrangement. As a matter of fact, in the era of electric locomotive evolution, the PRR engineers tried very deliberately to use steam locomotive underframe designs to a very large extent when planning the electric loco counterpart. Even stranger still is the fact that the early Baldwin diesel designs stole underframe designs from the electrics. Hence the Baldwin Centipedes, for example, strangely resembled GG-1 underframes! Thus, a DD-1 was in reality two 4-4-0's coupled back to back. If the Pennsy had ever built 4-4-4 steam locos, they could have been the Class O-1 and the electric locos would have been the O-2, assuming the steamers had been built first.

Therefore, for the purposes of this book and easy assimilation of the electrics to their steam loco counterparts, we shall refer to the loco arrangement as 4-6-4, 2-6-6-2, etc. rather than the less familiar 2-C+C-2 type designations.

The photo *(below)* illustrates the 2-8-2 wheel arrangement of L-6 loco 4790, the former 5938 prior to the 1966 Penn Central merger re-numbering. The unit was built by PRR shopmen at Altoona as the 7825 in 1932 and quickly re-numbered 5938 in 1933. It is seen here at Sunnyside, March 25, 1967.

(Al Roberts)

The DD-1
Where it all began

There were a total of 66 DD-1 units built (33 pairs) in the 1910-11 period. All were built at Altoona Shops with Westinghouse electrical equipment. All were D.C. third-rail powered and were designed and built in conjunction with the late-1910 opening of Penn Station, New York. In 1924, a replacement for the DD-1's was designed and built in the form of the L-5. As a result, 46 (23 pairs) were sold to the Long Island and those that remained on the PRR were retired in the late 1940's. That is, all but two pairs, the 3936-37 and the 3966-67 which were retained to pull the wreck derrick, the wire train, and do work for the phone company at night when the overhead catenary had to be de-energized to allow access to the telephone cables. Interestingly enough the L-5's were found less satisfactory than the DD-1's they replaced and they were themselves scrapped in 1942. Thus 20 of the DD-1's outlasted their intended successors!

In the photo *(above)*, the 3966-67 rests in its usual place at Sunnyside Enginehouse in March 1962. By October 1962 it would be severely cannibalized in order to keep its sister loco in operation a few more years.

The 3936-37 survived the 1966 pre-Penn Central renumbering, becoming the 4780-81 as shown *(below)* at the same location November 29, 1968. The Penn Central replaced the venerable veteran, which had outlived the Pennsy itself by almost a year, with an ex-New York Central motor. Ultimately this loco was enshrined at Strasburg where it truly belongs.

(Above, Matt Herson, Volkmer collection, below, Ken Douglas, Sweetland collection)

The B-1 Yard Goats

The B-1 electrics were built in two groups. The 3900-3901, 3910-3921, which were originally operated in pairs between 1926 when they were built and the mid-1930's when they were broken up into single units. A second group built later in 1934, 5684-5697 were always operated as single units. All were built in Altoona Works. The 3900's were Westinghouse equipped and the 5680's Allis-Chalmers equipped. Operation of some B-1's at Sunnyside Yard lasted into the 1966 renumbering. Numbers 3910, 12, 13, and 19 became 4750-53. Numbers 5685, 87, 90, 93 and 94 became 4754-4758. One unit, the 4756 (ex-5690) has been preserved at the Pennsylvania State Railroad Museum at Strasburg, Pa.

(Above) A low angle sun shows the detail well on the 3900, the alpha of the B-1 fleet, built in 1926. Having run its life out doing switching duty at 30th Street Station and Penn Coach Yard, it sits in retirement in 1960 at Philadelphia's 46th Street Enginehouse.

The 5687 *(opposite page, top)* was hauled out of retirement at Enola roundhouse and given a good cleaning and painting. On September 13, 1962 it was dispatched light (15 mph max!) to its new home at Sunnyside Yard. All B-1 operation at Harrisburg passenger station had been discontinued the previous year with the demise of a large portion of the east-west passenger train fleet.

(Opposite page, bottom) After the 1966 renumbering and into the Penn Central era, the 4751 cools its heels at Sunnyside Enginehouse on November 29, 1968. It was formerly numbered 3912.

(Above- Marty Zak, Volkmer collection, opposite page, both- Bill Volkmer)

The only L-6a

In 1930, Pennsy contracted with Lima Locomotive Works to build the carbodies and running gear for thirty 2-8-2 electric freight locomotives similar to the 7825-7826 (later renumbered 5938-5939, and still later 4790) which had been built in the Altoona Works. Electrical components were to have been applied at Altoona on delivery of the carbodies. The two Pennsy-built prototypes were class L-6 and because the Limas had certain minor differences giving them higher tractive effort, they were to be classed L-6a. One Lima unit, 5940, had been delivered and was in service in 1933 when the railroad discovered that the P-5's were a total flop in passenger service and were to be relegated to the freight service that the L-6a's had been intended for. So for the next 34 years, the 5940 plied back and forth between Sunnyside and Penn Station doing yard switching work. The 29 unfinished units sat at Altoona until 1942 when they were converted to B-29 bombers for the war effort. The 5940 was renumbered in the 1966 renumbering to 4791 but did not see service after merger day, being scrapped in late 1968.

(Above) 5940 is between engagements at Sunnyside Enginehouse, March 15, 1960.
(Below) 4791 (2nd) is enroute to Hollidaysburg for a date with the scrapper. Seen at Morrisville, August 27, 1967. *(Above- Bill Volkmer, below- Ken Douglas, Volkmer collection)*

The O-1's

Prelude to the P-5's

A total of eight 4-4-4 wheel arrangement electric locos were built at Altoona in 1930-31. They were built in pairs as follows: 7850-51 class O-1, 7852-53 class O-1a, 7854-55 class O-1b, 7856-57 class O-1c. Because each pair was a little different electrically, they could only MU by class (i.e. with each other). Their high drivers and lack of weight on the drivers rendered them as "slippery" and PRR quickly abandoned this wheel arrangement in favor of the P-5's. Six of the eight O-1's were scrapped in the late 1940's. The 7853 and 7857 survived into the early 1960's as transfer engines between Sunnyside Yard and Penn Station by virtue of their being steam boiler equipped, a quality not shared with their step-brothers and sisters, the L-6's and B-1's. The 7853 *(above)* class O-1a is resting at the Sunnyside Enginehouse in the early morning sun of February 23, 1960. At the time this portrait was taken, she had only about a year of life remaining. *(Bill Volkmer)*

The P-5's

"Making do" for 35 years!

The P-5's were built essentially in three groups, if one counts the two prototypes 4700 and 4791, the boxcabs 4701-4742, 4755-4774, and the modifieds 4743-4754, 4775-4790. One might make the observation that prior to the P-5's, Pennsy only made mistakes one or two at a time. This time they made 92 mistakes! Probably the most notable problem with the P-5's was the severe lateral thrust forces which the locomotives experienced. This tore up both the locomotives and the track structure over which they ran. In deference to the Pennsy engineers of the early 1930's one should remember that when being tested, the P-5's were being compared to the K-4 steam locomotives they were intended to replace. The P-5's were not exactly sensational on the grade coming out of the Bergen Hill Tunnel, but they beat the socks off the K-4's on downtime, ease on the fireman's back, emissions, etc. so they were judged overall to be a success. It was when the wires were extended south to Wilmington that the P-5's showed their inability to sustain 75 mph with heavyweight passenger trains, leading to the development of the GG-1.

In the 1960's, the deferred-maintenance P-5's operating on the deferred-maintenance trackwork caused many derailments. The center drivers on the P-5's were blind (i.e. no flange) and if the lateral thrust bearing plates (leg liners) were worn beyond limits, the center driver would drop to the crossties quite frequently. Lack of maintenance on the running gear would also cause the rubber drive-cups to self destruct in a matter of days and the P-5's went through drive-cups like popcorn. The GG-1's, by comparison, were relatively easy on drive-cup consumption.

During the last days of P-5 operation, the old maintenance hands at Enola had loads of methods of band-aiding the old girls down the road. The trucks and running gear were painted over about once per month to hide the cracks. The shop carpenter spent hours on the roofs spreading roofing compound around the headlight casings because rain would drip down the engineman's and fireman's (and brakeman's) necks! Probably the niftiest "smoke-and-mirrors" maintenance trick of all was when the truck equalizer fulcrum pins would break and the outer half of the pin would fall out onto the right-of-way. Rather than put the loco on the drop table in the house to replace the entire pin, the mechanics would get "one more trip" out of the unit by hammering a flashlight battery into the hole left by the missing pin piece and then smearing grease over it! To this day, whenever I ride in an airplane, I wonder if somewhere out there there is a Hanger Foreman with a flashlight battery and some grease telling the airplane mechanics "this'll get one more trip out of this old girl." Or, I wonder, did they paint the fuselage in order to hide the cracks?

(Above) P-5a 4740 rests beside its maintenance point, Enola roundhouse in September 1961. It was one of 25 P-5's rehabbed for continued service while the E-44's were being delivered. *(Bill Woelfer, R.S. Short collection)*

The only P-5, 4700

During the post-steam era, Pennsy management was death on fantrips. The top brass harbored much ill-will towards the fans after some disastrous delays were created during the last K-4 trips on the NY&LB in the 1956-57 era. Besides, there were not nearly enough extra coaches to go around for regular service as it was. So the author was amazed when in April 1961 he found out that the PRR management had actually authorized a fantrip using, of all things, a P-5 to run from New York City to Lancaster and return. I was equally surprised when I personally was asked to pick out the loco for the trip. My boss simply directed: "Pick out a dependable P-5 out of the bunch for your friends to ride behind next week." I replied that there is only ONE "P-5," the 4700! Without even looking up at me he muttered "whatever." So, out came the paint gun and the 4700 got a fresh coat of Brunswick green and on Friday, April 28, 1962, it was dispatched as a lead unit on a coal train to South Amboy. Then on Sunday, April 30th Al Holtz caught the special *(above)* at Langhorne on the Trenton Cutoff and Bob Watson, one of my Pennsy bosses in later years, caught it at Lancaster *(below)*, about to begin the return trip. This was the one and only time a P-5 of any description soloed on a passenger train during the last several years of P-5 operation. The 4700 was originally built as the 7898, along with sister 7899, later renumbered 4791. It (they) differed externally from the production P-5a's in its marker light arrangement. Number 4700 is also the only preserved P-5, it now reposing at the St. Louis Museum of Transport. What a shame it isn't at Strasburg with the rest of the PRR collection. *(Above- Al Holtz, below- Bob Watson)*

The P-5b

Throwing good money after bad

In 1937, the Pennsy management began groping around for ways to improve the tractive performance of the P-5a in freight service. After five years it was painfully evident that the P-5a would remain the electric freight locomotive of record on the railroad until they all wore out, or money could be found to design, test, and build a suitable replacement for them.

With this in mind, the 4702 was modified by adding traction motors to the engine trucks and additional air intake louvres to the carbody in order to improve the ventilation of the traction motors and other electrical apparatus on board. It is reported that the traction motors added were very similar to the motors then being applied to the new MU cars under construction, the bride and grooms. Very little was apparently written about the P-5b experiment, but it is obvious that the engineering staff early-on concluded that the added cost of conversion amounted to throwing good money after bad because one of the basic faults with the P-5a was its poor riding qualities, which could not be helped by upping the horsepower!

The only places that the author saw the 4702 operate was as the Baltimore tunnel helper and as the power for the Wilmington to South Philadelphia turn, places where it could be used interchangably with the equally ailing DD-2, 5800. One railroad employee told me that the 4702 posessed MU car-type controllers which could explain why it was never seen multipled with other P-5's. The 4702 was scrapped along with the 4790 in 1960, the first P-5's to go. The photo *(above)* shows the 4702 out of service in Wilmington in January 1960.

(Les Broomfield, R.S. Short collection)

The P-5a's

They came in two varieties

After the P-5a boxcab locos were built and delivered in 1931-32 (4701-4742 and 4755-4774), the 4772 was demolished in a grade crossing accident. As a result, the remaining undelivered units (4743-4754 and 4775-4790) were quickly redesigned to resemble the GG-1 styling, but with a riveted aluminum steeple-cab carbody. Most of the interior electrical gear was identical but rearranged to make room for the cab and engine crew.

Electricians were required to be of the tiny variety in order to work in the close confines of the noses of the modifieds. The two accompanying photos illustrate the similarities of these two groups and also their differences. In the photo *(above)* 4769 exemplifies the P-5a boxcab configuration.

(Below) This can be contrasted with the P-5a modified version illustrated by the 4746 running with the 4736-4726 as they head towards South Philadelphia yard in the vicinity of Girard Point yard lead on March 29, 1962.
(Both- Bill Volkmer)

Rough Riders All

When the author was dispatching/assigning locomotives eastbound at Enola, he came to know each of the locomotives and their crews like so many children and their toys. Certain "children" (crews) like certain "toys" (locos). For instance the 4800 was too cold in the cab and the E-44's had nice warm cabs, but their high center-of-gravity caused them to rock violently on jointed, rough track. The G's could roll a train fast but had poor starting torque, a quality badly needed for starting heavy tonnage, especially on grades. The P-5a boxcabs rode like baby carriages but their cabs leaked profusely in rainstorms! In the days of steam locomotives, the foreman used to keep an M-1 with a short tank fired up and headed east on the ready track beside the P-5's. When a crew coming on duty would complain about being assigned a P-5 instead of a G sitting on the next track over, the foreman would simply offer the fireman the M-1 in lieu of the P-5, meaning he, the fireman, would have to do some work for a change. That changed their attitude real quick! The GG-1's were horribly hot in the summertime and the P-5 modifieds were rough riders because the crew sat directly above the drivers. My boss always insisted that the modifieds lead the boxcabs, a rule never violated on eastbounds for crew safety in collisions. So, I used to say it was a lot like selling insurance, "some buy, and some don't!"

(Above) A typical P-5a set lead by modified 4776 accompanies 4756 and 4734 east out of Harrisburg Yard with a South Philly coal extra February 28, 1960.

(Opposite page, top) The 4412 departs Enola past Day Tower over very poorly maintained rail on her maiden voyage east, September 16, 1961.

(Opposite page, bottom) Modified 4781 whines over the Trenton Cut-Off near Willow Grove, Pa., in June 1961.

(Above and opposite page, top- Bill Volkmer, opposite page, bottom- Les Broomfield, R.S. Short collection)

The GG-1

World's most recognizable locomotive

The GG-1 needs no introduction to most railfans. People of the 1940's and '50's who were even only remotely aware of railroading in general recognized the streamlined GG-1 as the very symbol of the *Standard Railroad of the World,* the Pennsylvania.

No corporation going, railroad or otherwise had an image as indelible as the Pennsy's GG-1. The railroad used it everywhere from coach seat headrests, to magazine ads to timetable covers. Whenever Hollywood made a movie pertaining to travel on the eastern seaboard, they didn't use a B&O Pacific, they used a GG-1. When Lionel wanted to sell trains, the GG-1 was a sure fire winner. The nuts-and-bolts descriptions of how Raymond Loewy designed it and it was built etc. has been recorded in countless magazine articles and books. Therefore, we shall not attempt to recount it here. Rather, we'd like to perhaps look at the GG-1 in a different perspective. That is, how it lived life in the later years, after it had a few million miles under its belt.

The GG-1 birth in 1934 represented a major breakthrough in electric locomotive technology. After nearly forty rather frustrating years of experimentation, the PRR finally had come up with a winner. Most of the trial and error development in steam had been one or two locos at a time. With 92 trial and error P-5's on the road, heavily into the error column, the PRR engineers had to be running scared! So when the R-1 (4899) 4-8-4 versus the GG-1 (4800) 4-6-6-4 time trials were concluded and the GG-1 was declared the victor, the boys in Altoona had to be breathing a little easier. They had finally come up with a real winner. Or should we rightfully give the credit to General Electric? After all it was G.E. who produced the New Haven loco that the GG-1 was patterned after! The GG-1 was a great passenger engine. No one could deny it. For the eastern portion of the Pennsy, it was precisely what the doctor ordered. The only problem was the passenger trains they hauled were fast becoming victims of obsolesence. So sizeable quantities of GG-1's were rapidly becoming surplus. The P-5's were not eminently successful in freight service, but because of the fact there were 92 of them and the L-6a Lima 2-8-2 order had been cancelled, the Pennsy had to get along until a suitable freight electric loco could be developed (and afforded).

Because money was tight after World War II, the obvious choice was to regear 57 GG-1's (4801-4857) from 100 mph gears to 90 mph gearing and assign them to freight service. Their steam boilers were kept intact and they were equipped with train control and 90 mph was good enough for heavy mail trains. Thus every Christmas season the freight G's would temporarily go back into passenger service. In reality, there was no conscious effort to keep the freight G's off, for example, the Clockers, but a freight G on the BROADWAY was kind of a rarity.

The GG-1's did a good job on light-weight roller bearing-equipped *Truc-Train* traffic, but they left a lot to be desired on heavy-drag freights. The GE E-44, incidentally, with its dynamic-braking capabilities, ultimately becoming the Pennsy's only truly successful electric freight locomotive, was merely a clone of its diesel brethern. E-44's had only one real shortcoming, a shortcoming they shared in common with the GG-1's and P-5a's. They couldn't operate west of Harrisburg! Equally at home on fast passenger or fast freight, the GG-1 was always a fasinating sight to see.

(Opposite page, above) The 4928 rolls into 30th Street Station Philadelphia with Train 111, the PRESIDENT in post-merger June 1969. The trailing G has already been repainted Penn Central. *(Opposite page, below)* A classic GG-1 pose in its original five stripe paint without the snow screen added in later years. The 4824 is at Harrisburg passenger station in 1949.
(Opposite page, above- Bill Brennan, below- Nelson Bowers)

Old Rivets,
and converse

Any student of the GG-1 knows that the 4800, the original loco of its class, was known as *Old Rivets* because its carbody was entirely riveted and the following 138 locos were built with all-welded construction. The author, when working at Enola, often praised this decision because the 4800 was a tough motor to sell to the crews in the winter because the carbody allowed copious quantities of cold air in, but alas, the welded jobs were not an awful lot warmer. I remember one cold day the crew assigned to 4800 complained that the blowers were sucking all the cab heat out through the bulkhead door to the blower room. I took a roll of masking tape into the cab and completely masked the door shut and as I exited the cab down the ladder, I turned and tossed the rest of the roll of tape to the engineer!

What is seldom remarked in the railfan world is that the only non-riveted (i.e. welded) P-5a modified was the 4770 which received its welded carbody in January 1945 after the original boxcab was wrecked in an accident. Incidentally, the 4770 also became distinguished as the very last modified to go to the scrapper, in October 1962.

(Above) The 4800 has just emerged from the backshop for a test run following major surgery which removed her steam boiler and forever banished it to freight service. Date is June 10, 1960.
(Opposite page, top) The 4800 awaits a freight call at Enola pit, April 28, 1962.

(Opposite page, bottom) The 4770 is living on borrowed time as the last remaining P-5a modified. Seen here at Enola pit, June 16, 1962. Why was the running gear freshly painted? To hide the cracks in the frame and running gear from the Federal Man, that's why!

(Above and opposite page, bottom- Bill Volkmer, opposite page, top- Ken Douglas)

ALTOONA WORKS
6-42 GG1.4385

4800

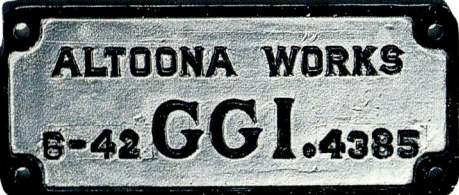

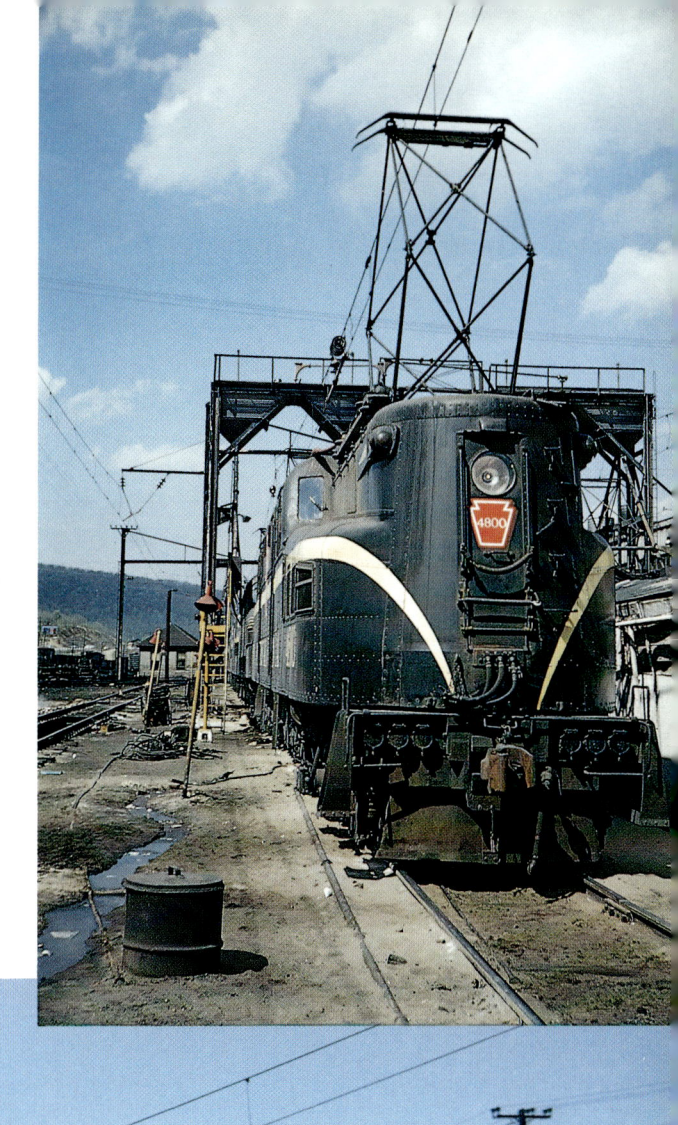

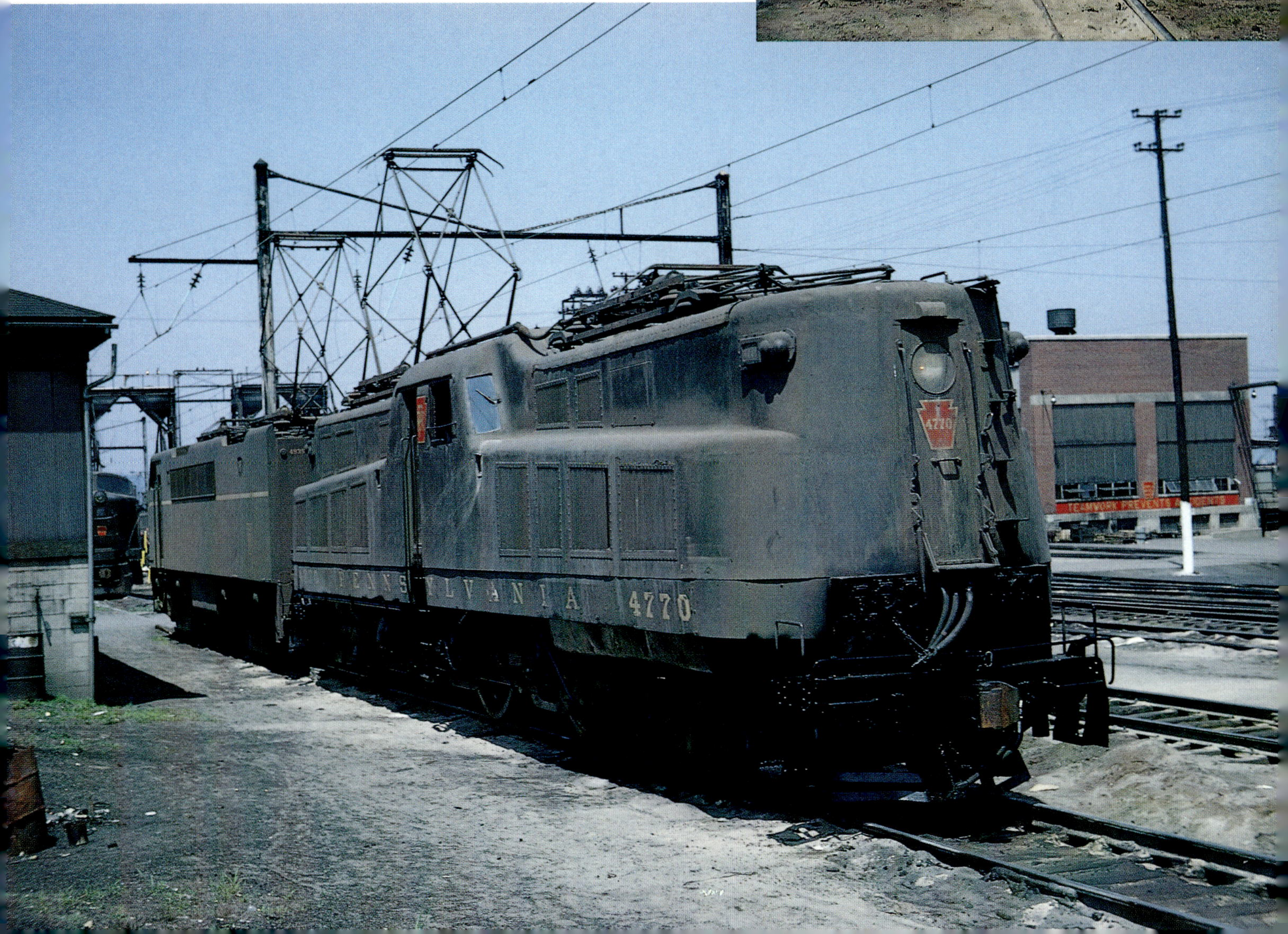
4770
PENNSYLVANIA 4770
TEAMWORK PREVENTS

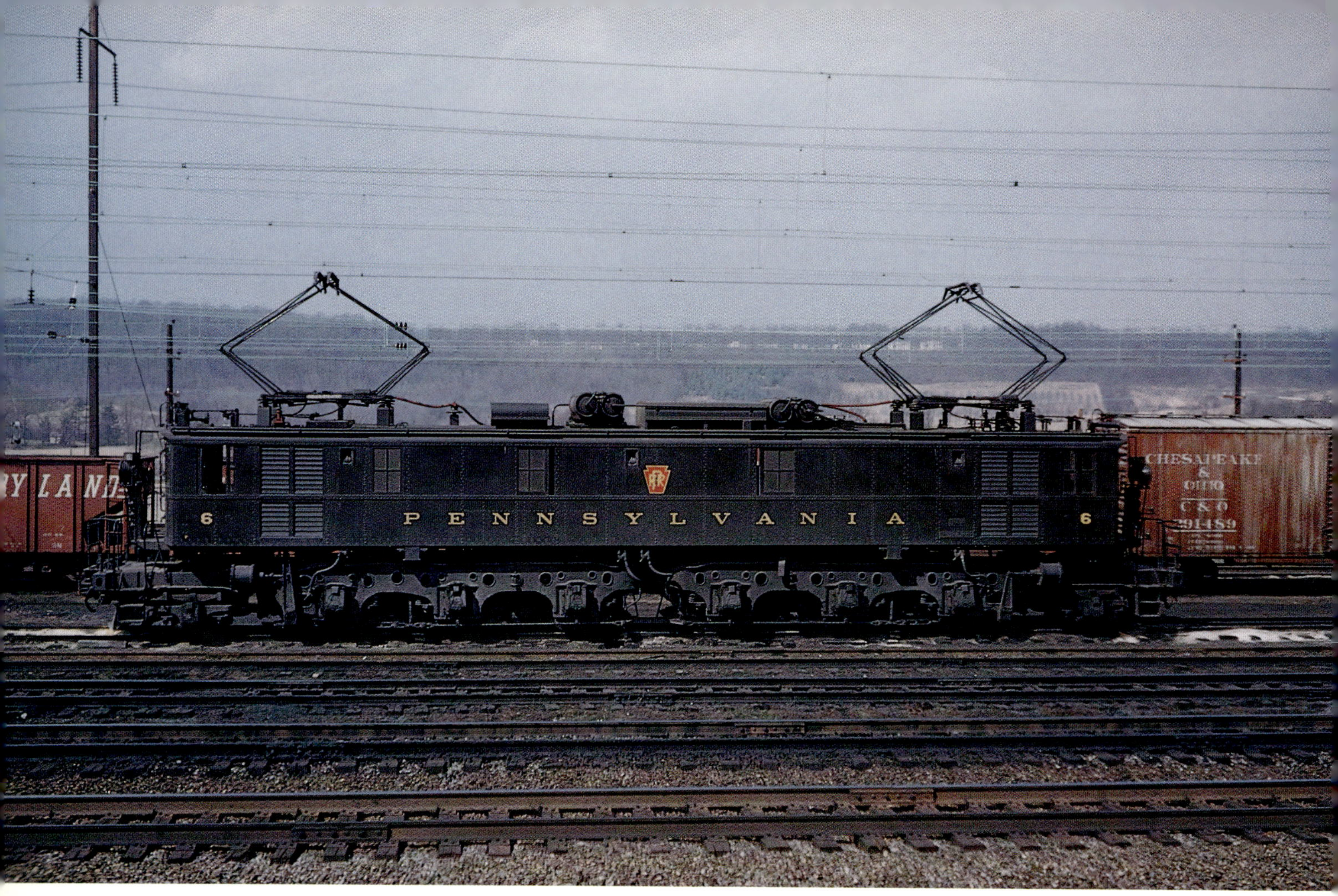
6
PENNSYLVANIA
6
CHESAPEAKE
&
OHIO
C&O

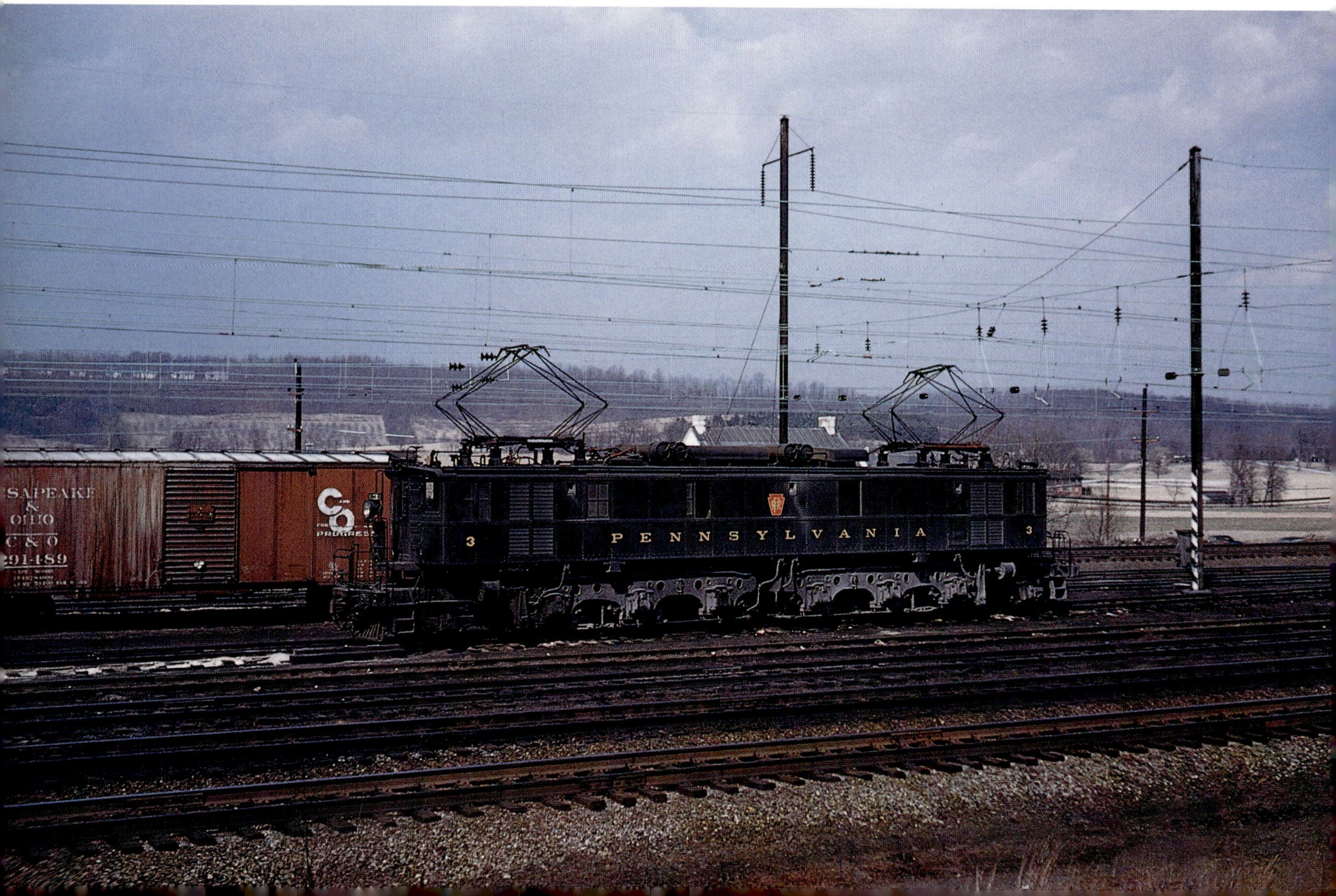
PENNSYLVANIA
3
3
PROGRESS

The FF-2's

Rolling Dynamos

In 1957, Pennsy was strapped for motive power and a suitable design for new purchases had not been finalized. After considerable shopping around, the road settled upon buying eight motor-generator driven locomotives which had been rendered surplus by the Great Northern's dieselization of the Cascade Tunnel.

The units were renumbered 1-7 and the eighth unit, the one with the EMD FT style cabs was stripped for parts at Altoona. With all the experimentation PRR had done over the years utilizing A.C. propulsion, D.C. propulsion, ignitron rectifiers and the like, the FF-2's certainly added a new dimension to the PRR motive power mix. Heretofore only one m.g. loco had ever been tried out and that was 50 years before! The FF-2's used A.C. line voltage to drive a huge motor which, in turn, rotated a large D.C. generator.

The D.C. generator powered six pairs of drivers with unbelievably high starting tractive effort coupled with a very low gear ratio. The net result of all this was that the Pennsy enginemen, being quite used to the GG-1's simply couldn't adapt to the FF-2's and would spin the wheels right out of their tires! We actually had to weld straps on the inside of the tires to keep them from falling off the wheels! Both pantographs had to be kept against the wire, lest a single pan would bounce away from the wire and the motor would get out of sync with the line voltage. Top speed because of low gearing on these FF-2's was 35 mph. At the crest of the grade, frequently the hauling GG-1's or P-5's would be in excess of 35 and the FF-2 could not gain enough speed to bunch the slack for the conductor riding the caboose to pull the pin to uncouple! By early 1961 all the FF-2's had been retired and put into dead storage at Enola roundhouse where they had been maintained. *(Opposite page, top)* A full broadside photo of number 6 at Thorndale helper station shows what a nicely proportioned unit the FF-2's were. March 21, 1959. Note that Pennsy had removed the side doors and ladder during conversion to PRR standards. *(Opposite page, bottom)* The 3 spot is shown in a 3/4 angle also at Thorndale. *(Below)* The 7 is also found at the traditional Thorndale resting spot April 15, 1960. *(All- Bill Volkmer)*

The E-2b's

General Electric Experimentals

In 1951 a replacement for the P-5's was desperately needed to haul the Eastern Region freight. General Electric built four class E-2b locos (numbered 4939-4942) in that year fashioned along the lines of the covered wagon road diesels of the day, albeit with larger diameter wheels to accomodate the necessarily larger A.C. traction motors. Even though they could multiple with the P-5's they were not considered a success largely because of the low available starting torque inherent in an A.C. motor as contrasted with the D.C. motors of the diesels of the day. G.E. also built a fifth and sixth unit as demos for the Great Northern, but alas GN didn't buy. Thus they became the PRR 4943 and 4944 two years later in 1953. *(Left)* The 4939 had just been scrubbed with bucket, brush and hose in the Enola roundhouse even though the unit was normally maintained at Wilmington Shop. Wilmington Shop seldom, if ever, washed a unit during Monthly Inspection, only when the unit was overhauled. Purpose of this wash job was so the photographer (me) could shoot a "clean" E-2b, there being no others on the railroad at the time. The date was September 13, 1962. The numberboard-over-the-headlight *(below)* on the 4944 exposes the unit's ex-demo heritage. Also this photo shows the extreme westernmost catenary on the PRR. Photo was taken February 17, 1960 at Harrisburg Yard. To the author's knowledge, this yard was the only point on the system utilizing wooden creosoted catenary poles. *(Both- Bill Volkmer)*

Westinghouse Rectifiers

In July 1949 Pennsy first experimented with ignitron rectifiers by putting one into the baggage compartment of MU car 4561 and substituting D.C. traction motors where A.C. once were. Convinced that the combination of D.C. motors and A.C. catenary was the wave of the future, Westinghouse Electric delivered four experimental locos to the Pennsy in the fall of 1951. Both Westinghouse and G.E. had been given more or less free reign to come up with the design of their choice to sell to the PRR. Two, the 4995-4996, were E-3b class (3 B-type trucks) and the other two, 4997-98, were E-2c class (2 C-type trucks). Note that in tribute to the age of dieselization, PRR had now discontinued steam classes for their electrics.

The ignitron rectifiers in the Westinghouse locos, being extensively water-cooled, turned out to be a plumber's nightmare. The Westinghouse engineers failed to realize that pipe wrenches have handles and those handles frequently require extensions to loosen rusted joints. There simply wasn't enough room to swing the handles on those wrenches! Also, water in cold weather needs anti-freeze which wasn't readily available when the water leaked out in winter and fresh water was added out on the road by a friendly Fire Department. Small wonder those units led a checkered life!

(Above) 4996, an E-3b rests on the inbound inspection track at Enola, June 11, 1962. *(Below)* This photo of the 4997 exemplifies the sister class E-2c, also taken at Enola, February 19, 1961. *(Both, Bill Volkmer)*

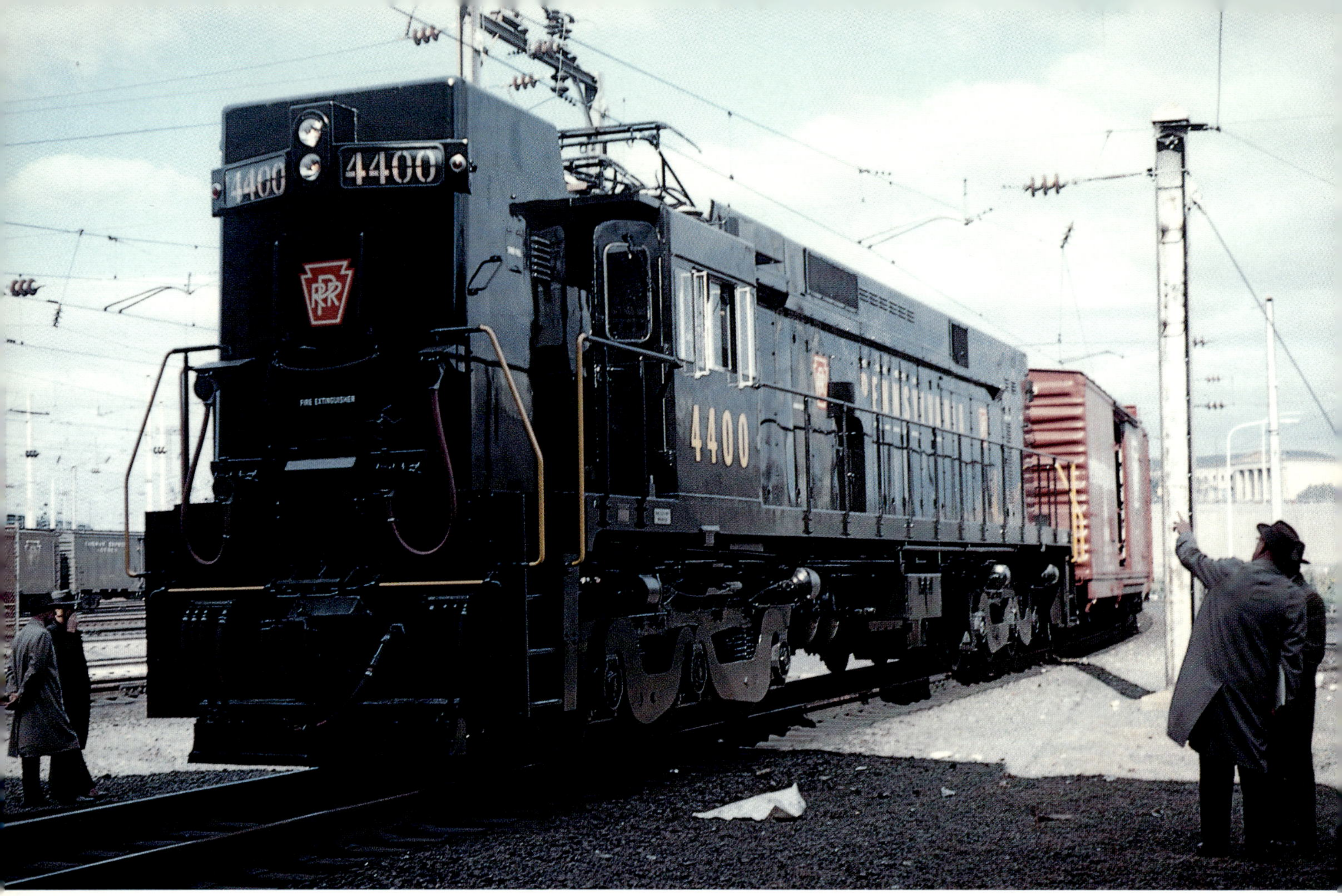
4400
4400
PRR
FIRE EXTINGUISHER
4400
PENNSYLVANIA

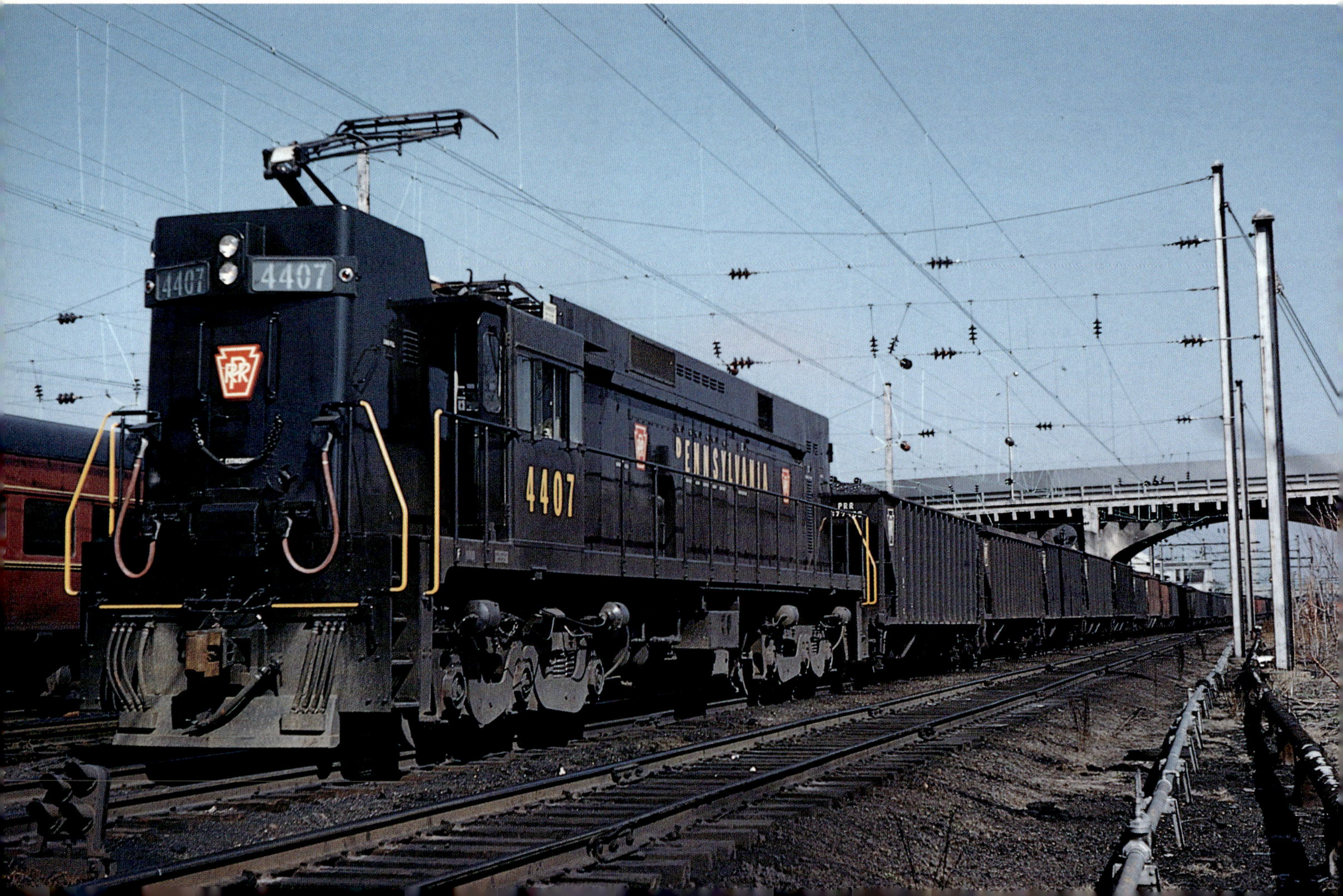
4407
4407
PRR
4407
PENNSYLVANIA

The E-44

The Ultimate Electric Locomotive

The evolution of electric locomotives suitable for freight hauling culminated in the 1956 creation of 12 rectifier locomotives built by General Electric for the Virginian Railroad. These locos were highly successful in not only hauling coal upgrade, but equally important, they were equipped with dynamic brake to bring the coal downgrade without destroying the locomotive wheels and brake shoes, a real problem with the GG-1's and P-5's.

By 1956 though, before considering buying new rectifier locos from G.E., the Pennsy had to cope with two severe problems. It was out of cash, having just purchased almost 2,465 diesel locomotives in 10 years, and the road's A.C. power generating equipment and associated infrastructure was beginning to show its age. PRR was torn between dieselizing the Eastern Region freight operation with the diesel's inherent added operating flexibility (routing bridge traffic up the Bel-Del to the L&HR and into New England around New York Harbor as an example) or upgrading the electric locomotive fleet through new purchases. The road took three years to make this very tough decision.

In 1959, the General Electric Co. made Pennsy an offer to lease 66 E-44 type locos on a fifteen year basis with an option to buy at the termination of the lease. They were numbered 4400-4465 and became the last locomotives to be delivered with the road's name spelled out in full on the flanks. In anticipation of the coming merger with the New York Central, all diesel deliveries from this point on were nameless on the sides.

The first E-44 was delivered in October 1962 and for the most part the locos were a total success, meeting or exceeding the road's expectations in almost all respects. Probably the most objectionable feature about them was their tendencey to rock rather violently on rough jointed track because of their abnormally high center of gravity. This was compensated for by their toasty warm cabs as far as the crews were concerned. As excellently as these locomotives performed, they still could not outshine their diesel counterparts when they were needed to run, for example, over the B&O around a derailment on the paralleling PRR Washington mainline!

(Opposite page, top) Possibly the very first photo ever taken of an E-44 under Pennsy wires, the 4400 was on display at 30th Street Station, October 25, 1960. A new X-29g box car was also on display.

(Opposite page, bottom) Brand new 4407 hauls an eastbound coal train through Harrisburg in April 1961. *(Above)* Also brand new E-44 4403 has been delivered to Enola pit on January 29, 1961. The pantograph had been raised to the wire for the first time just moments before this photo was taken, not even one inch had been registered on the odometer! Not counting test mileage at Erie and transit-in-tow to Enola.   *(All- Bill Volkmer)*

GG-1's and Snow

To see a GG-1 in snow was an eerie sight. A blanket of snow on the ground, especially around the noisy big cities of the East tends to put a damper on the ambient sound levels. So when a train hauled by a G approached in a snow storm, about all one could hear was an occasional wheel squeal! The G's tended to use a lot of sand anyway, even on dry rail, so on snowy rail it was business as usual as they went about plying their trade.

For the first 20 or so years of their life, the G's had no notable trouble coping with the rigors of snow. Then in February 1958 a very fine blowing snow storm did the G's in. A vast majority of the fleet suffered grounded out traction motors. A combination of Irish linen protective screens which had to be inserted each time snow was forecast and epoxy dipped armatures were fixes, but in later years a repeat disaster prompted PRR engineering to retrofit several GG-1's, beginning with the 4857, with a centrifugal dirt separator to get rid of the snow before it could be ingested into the traction motors. This accounts for the appearance of an extra "porthole" below the pantograph as seen on several GG-1's during their last years of operation.

(Above) When sleet conditions appeared, as they often did in late afternoon after an all day rain, "double pan" orders went up for all electrics so that the extra shoes would serve as sleet cutters. Such was the occasion with these GG-1's passing COLA Tower in Columbia, Pa., in the winter of 1968.

(Chet Fuhrman)

The Silver GG-1's

In 1952 when the stainless steel coaches, parlors, diners, and observations were purchased to re-equip the CONGRESSIONAL LIMITED, several GG-1's in the low 4900 series were repainted into the Tuscan Red scheme with the traditional five gold leaf stripe decor. The passenger diesel fleet which was then being augmented by the delivery of new E-8's was also given the Tuscan treatment.

In 1954, the railroad decided to hire a consultant to completely overhaul its managerial setup, reorganize the divisions into regions, and create a new visual image to the public.

This resulted in the hiring of an art director for the first time and he was given *carte blanche* to redecorate the dining car interiors, locomotive exteriors, the new Penn Center office building and everything in between including the annual report covers such as the 1957 one of Penn Center with "PRR" diagrammed in the office window lights!

The result of the visual redesign was the single wide stripe on the locomotives, the Keystones with 3-D shadows, fancy drapes in the parlor cars and psychedelic interiors in the 1908-vintage MP-54 MU cars! One might even say it inspired the dust jacket color for the book you are now reading, as well!

One noble venture was to paint three GG-1's silver with a red stripe to match the stainless steel passenger cars. Alas, the rusty water dripping from the steel pantograph shoes was too difficult to keep clean and the silver G's were quickly repainted back to Brunswick Green. Fortunately, one fantrip was run immediately after the 4866 was repainted in the silver scheme and Nelson Bowers photographed her *(above)* at Columbia, Pa., in what is believed to be the fall of 1955.

(Nelson Bowers)

#4460

Silicon Diode Rectifiers Introduced

As we noted earlier in this book, the downfall of the Westinghouse rectifier units of 1951 was the extensive liquid cooling system that the mercury arc, ignitron rectifiers required which were prone to leaks and freeze-ups. The early E-44's (4400-4459) were also water-cooled but had the benefit of close to ten years of engineering improvements to help alleviate the problems. About the time the E-44 deliveries were up to 4435 (April 1962) G.E. had decided that they had perfected the state-of-the-art of silicon rectifiers, which required no water-cooling, to the point where they could deliver a reliable test unit. So the 4460 was delivered in July 1962 out of sequence and it was classed E-44a. The builder's number indicates it rightfully should have been #4436. Each silicon diode rectifier was about the size of a pack of cigarettes as opposed to the ignitrons which were about 3 feet tall and roughly a foot in diameter. The advent of the silicon rectifier was solid state electronics coming of age in the railroad world. The year 1962, incidentally was the same year that the railroads were introduced to the GP-30's and U-25B's which were also heavily into solid state circuitry. The remaining E-44a's, 4461-4465 were delivered in sequence after 4459 in mid-1963. Later Pennsy retrofitted several E-44's starting with the 4459 and working backwards making them class E-44a also.

(Above) Shortly after it was put into service, 4460 is seen teamed up with 4436 and 4405 at Enola on train SP-8, a South Philadelphia hotshot containing mostly perishable commodities, past Day Tower on December 1, 1962.

(Opposite page, top) Shocks Mills Bridge was an almost impossible place to photograph because of lack of access and dense foliage. "Four fours," the 4444, which was never converted to silicon, is westbound to Enola in May 1969.

(Opposite page, bottom) Make no mistake about it, this is Pennsy. Position light signals, E-44's and snow! The 4437 wheels a hopper extra eastward on the Port Road, 3.5 miles east of Columbia, Pa., at Washington Boro on a snowy day in February 1968.

(Above- Bill Volkmer, opposite page, both- Chet Fuhrman)

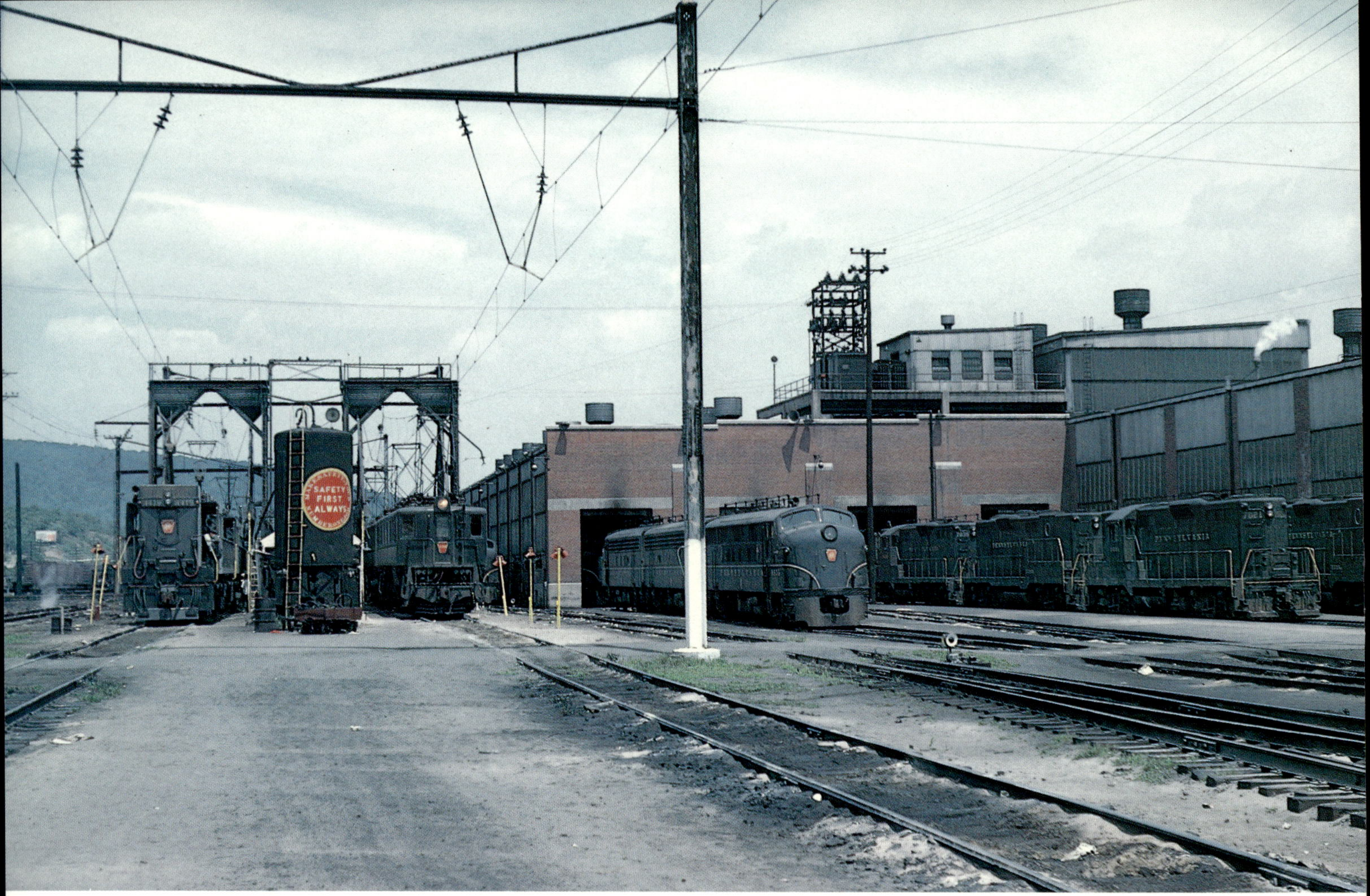
SAFETY FIRST ALWAYS
PENNSYLVANIA

PENNSYLV NIA 4759
PENNSYLVA A 9526

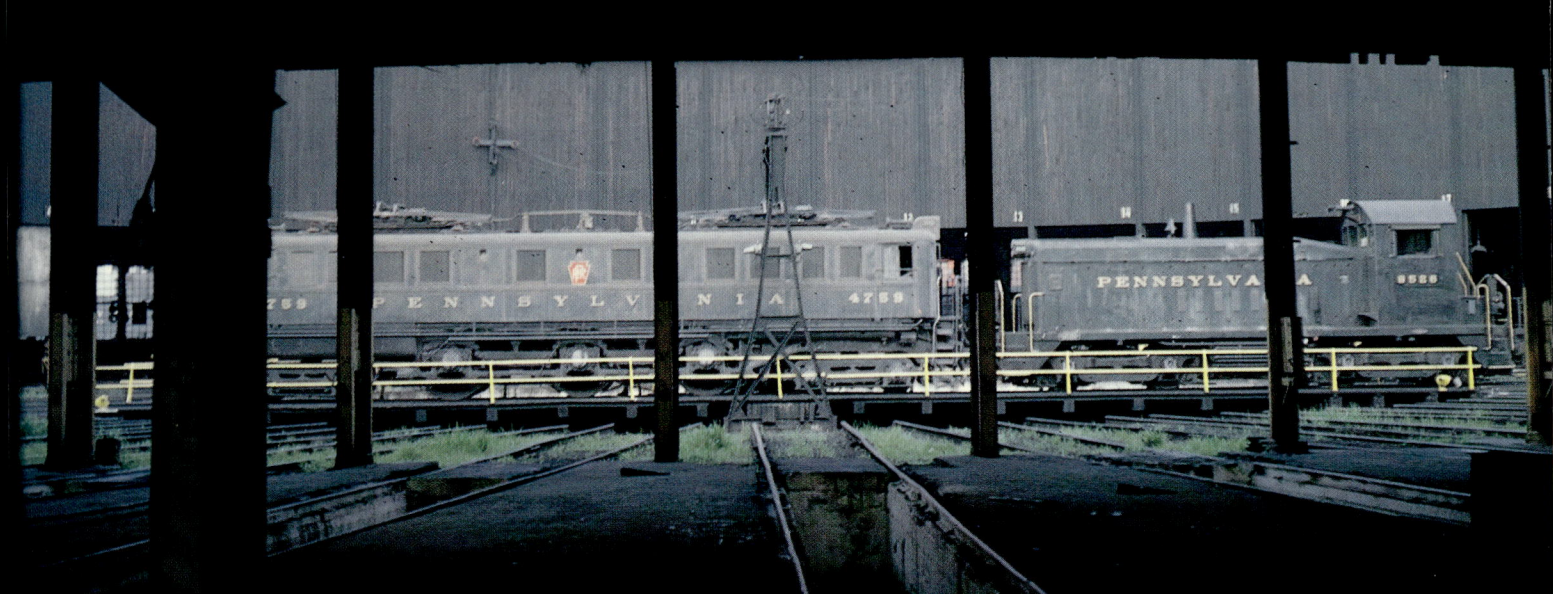

Enola

Hub of Freight Activity

Enola Yard was the gateway to the East, so to speak, on the Pennsy. The author had the good fortune to be assigned there as Assistant Enginehouse Foreman at the roundhouse in April 1962. For a year I worked the second trick having charge of the maintenance of those P-5's that remained on the roster. In April 1962, forty-six of the original ninety-two remained including the last three P-5a modifieds 4746, 4749 and 4770. E-44 deliveries were up in the 4430's. They were being delivered at the rate of about two per month and the P-5's were being withdrawn at about the same rate. The Enola roundhouse, not being adequately equipped to maintain roadswitcher type carbodied E-44's, was itself being phased out and the new E-44's were assigned to the Harrisburg Passenger Diesel Shop for maintenance.

Most of the day to day chores at Enola consisted of inspecting and sanding a myraid of GG-1's (in freight service), P-5's, and E-44's along with an occasional E-2b and E-3b. All the FF-2's and Harrisburg assigned B-1's by this time were in retirement, stored in the unused half of the 360 degree roundhouse. Serviced locos were then dispatched to various points such as Meadows, Greenville, South Amboy, South Philly, Baltimore, Pot Yard, Pavonia, and Earnest. Second trick was by far the busiest time of the day for eastbound dispatchments. Third trick was when the heavy westbound (i.e. inbound) movements occurred and first trick was when most Monthly Inspections were performed on the P-5's.

(Opposite page, top) Join us now for a tour of Enola electric pit and all the interesting sights to be seen there from day to day. We begin with an overall view of the electric pit and the Enola Diesel Shop. The track on the left was reserved for GG-1's and E-44's as they were first out for the trailer trains and other hot-shots. The track on the right sanded the P-5's used for the secondary trains. The diesels seen here are inbound to the diesel shop. Scene was taken July 22, 1962.

(Opposite page, bottom) The shop diesel switcher brings a P-5 onto the turntable inside the completely enclosed Enola roundhouse. This was a chore unique to electric locos wherein they could not operate under their own power into the house. July 7, 1962.

(Below) Boxy #4444 relaxes before beginning a lifetime of hardwork and her first trip eastbound on September 15, 1962.

(All- Bill Volkmer)

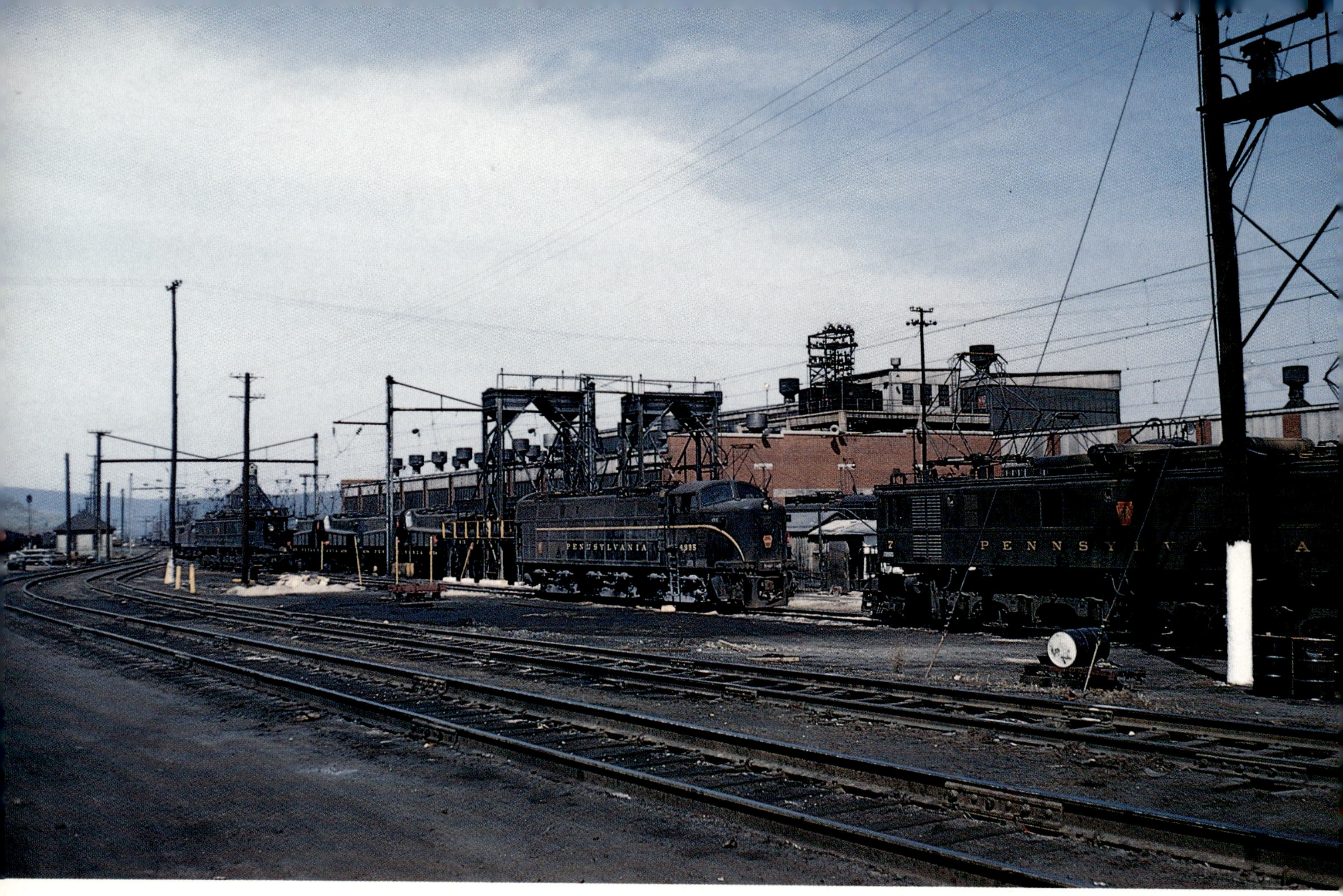
PENNSYLVANIA
PENNSYLVANIA

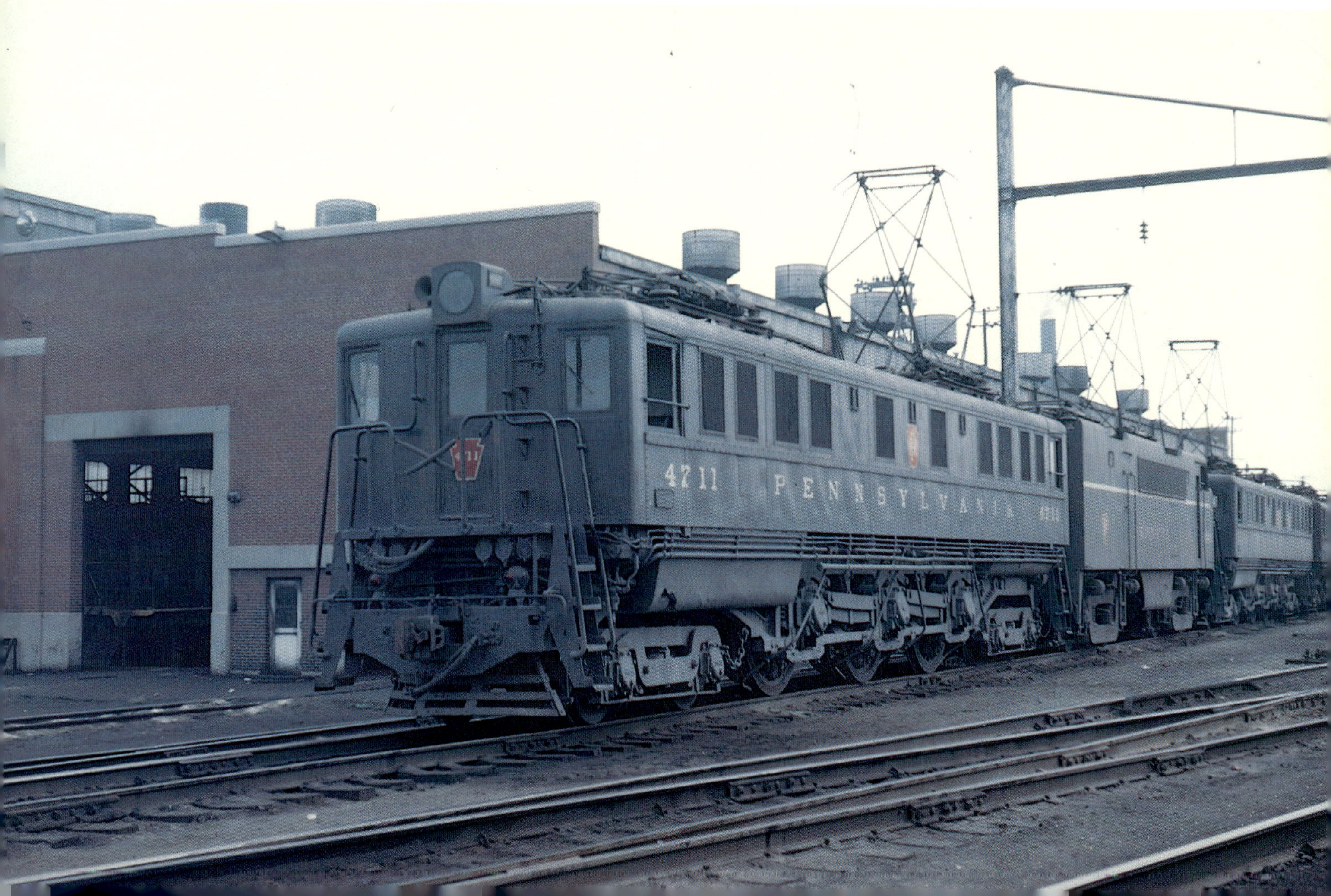
4711
PENNSYLVANIA
4711

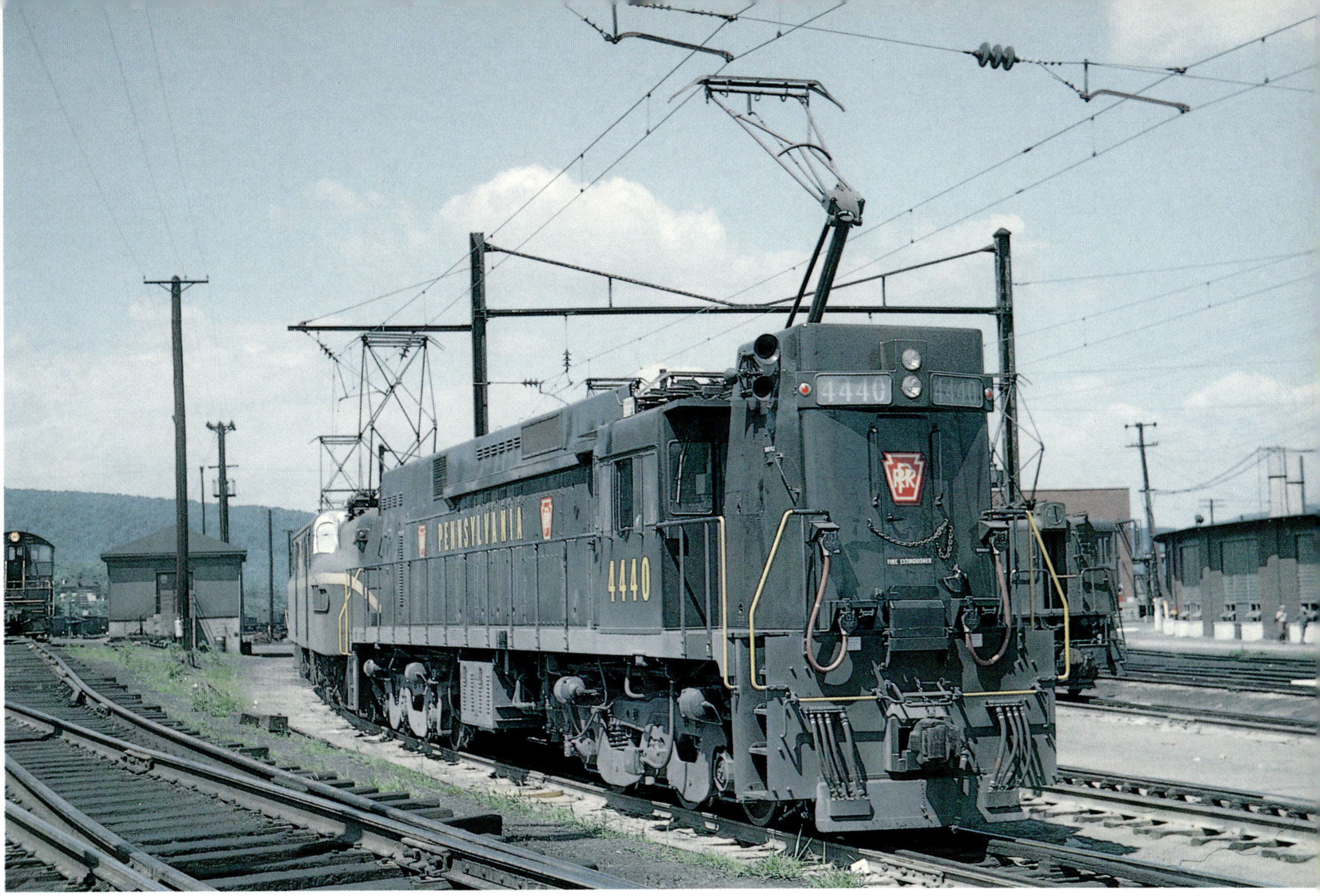

Enola Hump

Railfan Barrier

Railfans habitually like to sneak onto railroad property, grab that quick shot and disappear into the crowd. That could be done virtually at any roundhouse or engine terminal anywhere. Anywhere, that is, except Enola. The hump there was operated around the clock and a slowly moving endless flow of freight cars over the hump formed a natural barrier 'tween the fans and the motive power. Since the author served as resident railfan/Supervisor, and since the author had a folded up Kodak *Retina III* in his jacket pocket at all times, the hump presented "no problem."

(Above) The outbound ready track held the 4440 on a lazy July 4, 1962. The unit was still new and thus was relatively clean.

(Opposite page, top) The third pit track (closest the highway) was where the odd-ball locos were kept as their call to duty was only occasional. An E-3b and an FF-2 repose there on an April day in 1960.

(Opposite page, bottom) P-5a 4711 and an unidentified E-2b are beside the diesel shop August 5, 1962.

(Above and opposite page, bottom- Bill Volkmer, opposite page, above- Bob Watson)

Electric Locos
vs. Washers

In the entire electrified zone of the PRR, there was only one automatic locomotive washer equipped with overhead catenary. This was located at Sunnyside Yard on the turning loop and was designed to only wash GG-1's and coaches. Because of the proximity of the high voltage pantographs, even this was a tricky thing under the best of circumstances. Thus, the freight GG-1's and other Wilmington-based electrics just simply never got washed. Occasionally a freight G (4801-4857) would do a turn as a passenger loco and make a trip through the washer. Unfortunately, for long periods of time the washer could be out of service for repairs, then the GG-1 fleet would get to looking pretty grubby in no time flat!

The P-5's, thanks to the interested management at Enola where they were maintained, would get a bucket-and-brush bath once a month in the roundhouse where there was no overhead wire to interfere with the washing.

When the E-44's were delivered, they were assigned to the Harrisburg Diesel Shop and were never washed. The author, in the role of Assistant Enginehouse Foreman at Enola took it upon himself one day to grab six very crusty E-44's, drop their pans, bleed the air, and drag them through the diesel washer using a shop switcher. The result was six temporarily clean E-44's as seen *(above)* on July 18, 1962 and one very irate diesel house Foreman also seen in the photo. The total disruption this caper brought the inbound diesel flow to the fuel racks and inspection area caused the washing of E-44's never to be repeated here again!

(Opposite page, top) GG-1 4903 threads its train through the Sunnyside washer on March 31, 1968. After its external wash the coaches and sleepers will be cleaned on the yard storage tracks in time for a late afternoon departure.

(Opposite page, bottom) An example of a P-5a just out of the Enola "hand-wash is the 4742 on August 16, 1962. Compare to the still young, but dirty, 4436 to its rear!

(Opposite page, top- Al Roberts, others-B. Volkmer)

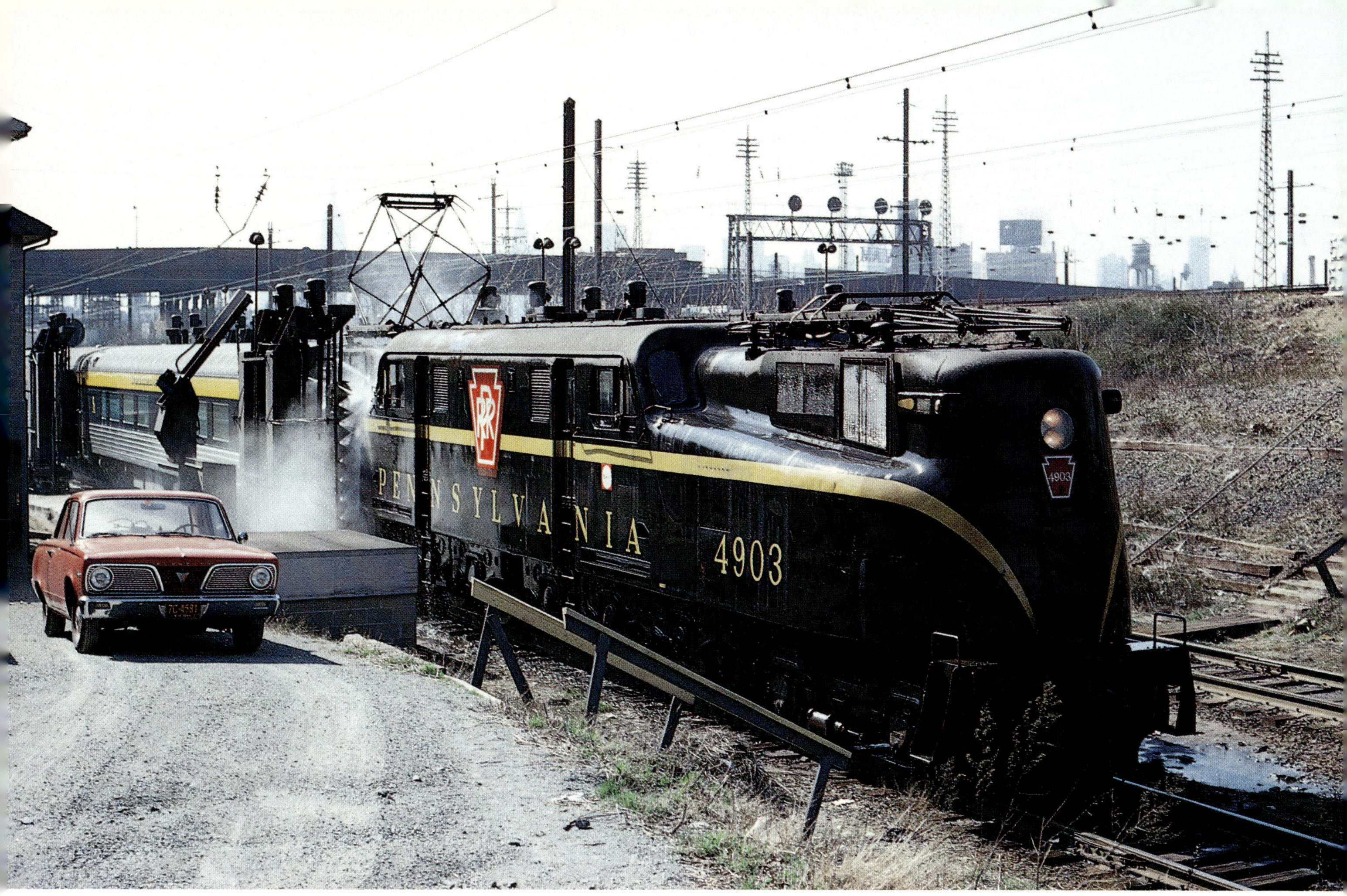
PENNSYLVANIA
4903

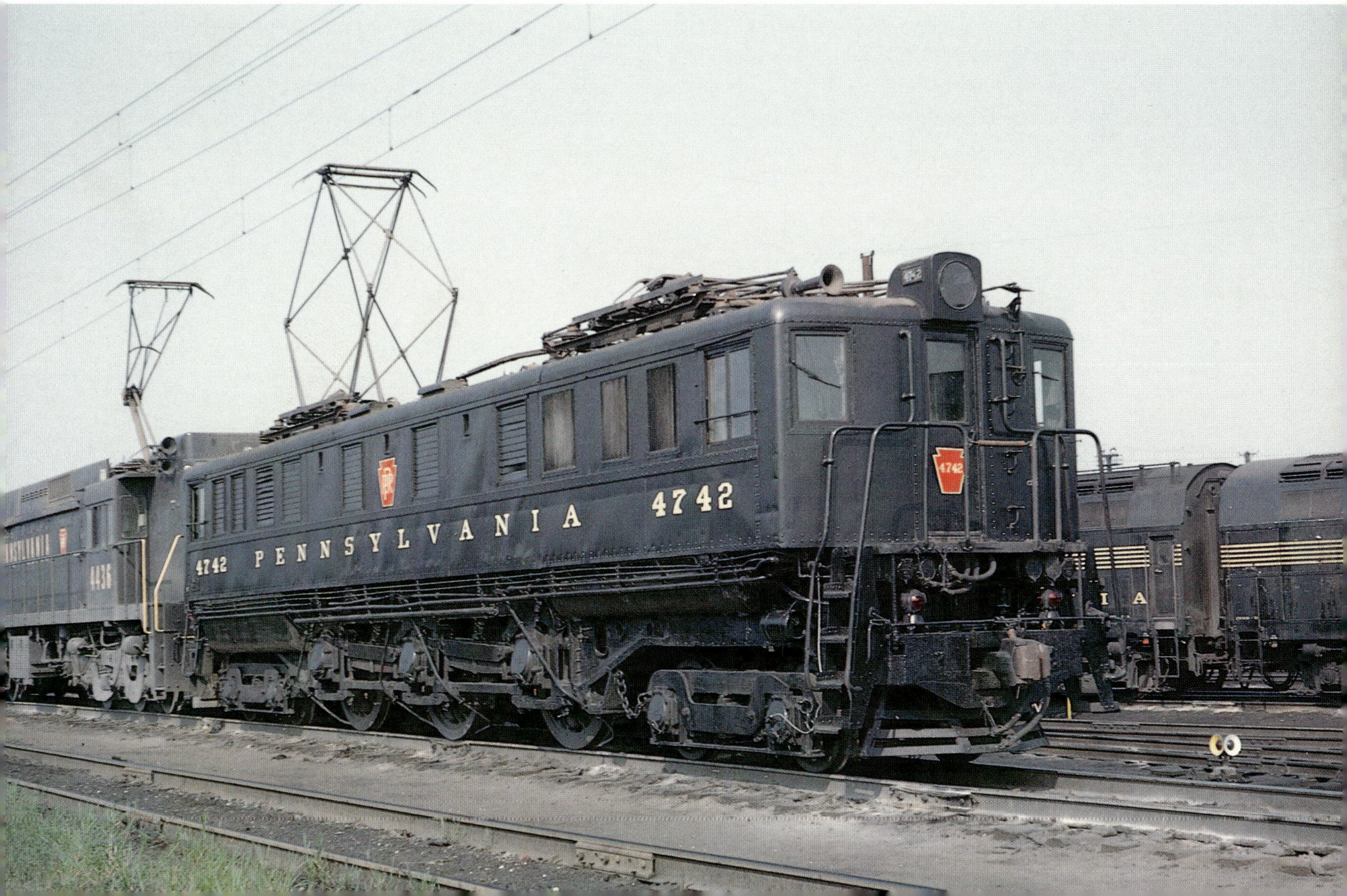
4742 PENNSYLVANIA 4742

The Leaning Tower of Lemo

Lemoyne, Pa., is a town just south of Enola and on the west shore of the Susquehanna immediately opposite downtown Harrisburg. It is where the junction of the Cumberland Valley Railroad, by now the PRR Hagerstown Branch and the Pennsy main freight line lay. PRR electrified the CV bridge over the Susquehanna for the sole purpose of ferrying motors (as the electric locos were called) to and from Harrisburg so that freights could bypass Enola and freight GG-1's could be drawn into passenger service in a wink. To the author's knowledge, Lemo, a tower which leaned as it slowly sunk into the ground towards the river, was the site of the only 90 degree electrified crossing on the Pennsy. The photo *(above)* illustrates how the tower leaned as the 4742-4783 clattered over the diamond eastbound. *(Below)* The freight it is passing this January 9, 1960 was a westbound with 4763-4737.

(Above) Not all the trains passed the tower at ground level. The two tracks nearest the river passed beneath the CV bridge as 4995-4997 are doing April 6, 1963. Turn the page for a view taken *from* the bridge shown in this photo.

(Below) In a Christmas Card-like setting, brand new E-44's 4401-4402 cross southbound on January 28, 1961 on the CV preparatory to wying and going to the pit at Enola. The duo had previously dropped their train at Harrisburg Yard. *(All- Bill Volkmer)*

New Cumberland

The town of New Cumberland, Pa., is little more than a wide spot in the road located on the west shore of the Susquehanna River immediately south of Lemoyne. North of New Cumberland, the four tracks spread into two pairs, one pair rising to the level of the Cumberland Valley Branch which crossed the main line at Lemo, the other two passing under the CV at Lemo.

(Above) This photo taken from the CVRR Bridge shows the grade separation described on the previous page. The two Westinghouse units are headed westbound (north) towards Enola on the river track, lower level while the two G's head east on a parallel track.

(Opposite page, top) On June 30, 1962, when this photo was taken, the P-5's were in their twilight as the 4738-4770 trudged westward towards Enola through an increasingly weed infested right-of-way.

(Opposite page, bottom) With the E-44's almost all delivered, the 4433-4455 whine eastward past the long unused New Cumberland passenger station.

(Above- Al Holtz, opposite page, both- Bill Volkmer)

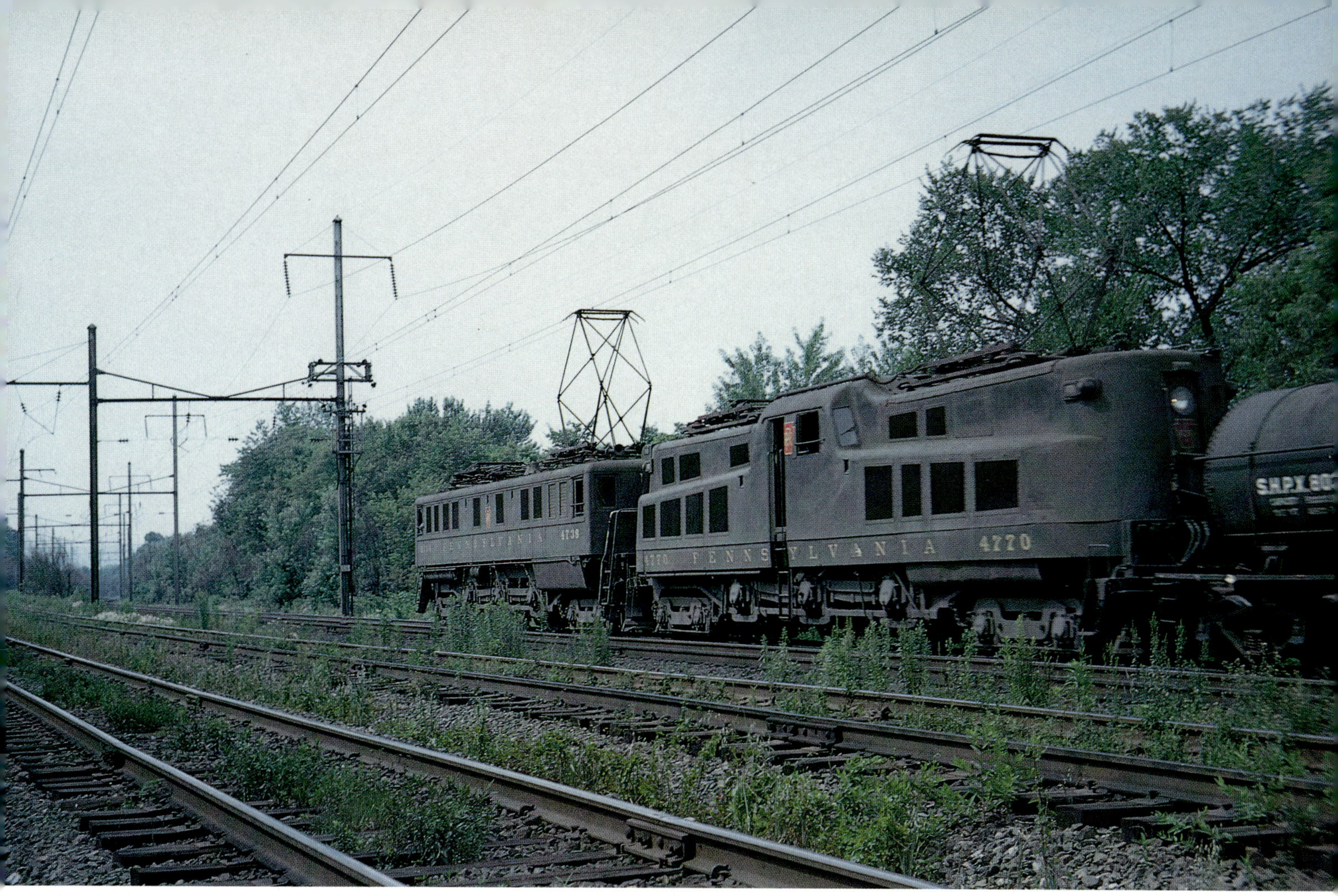
PENNSYLVANIA 4770

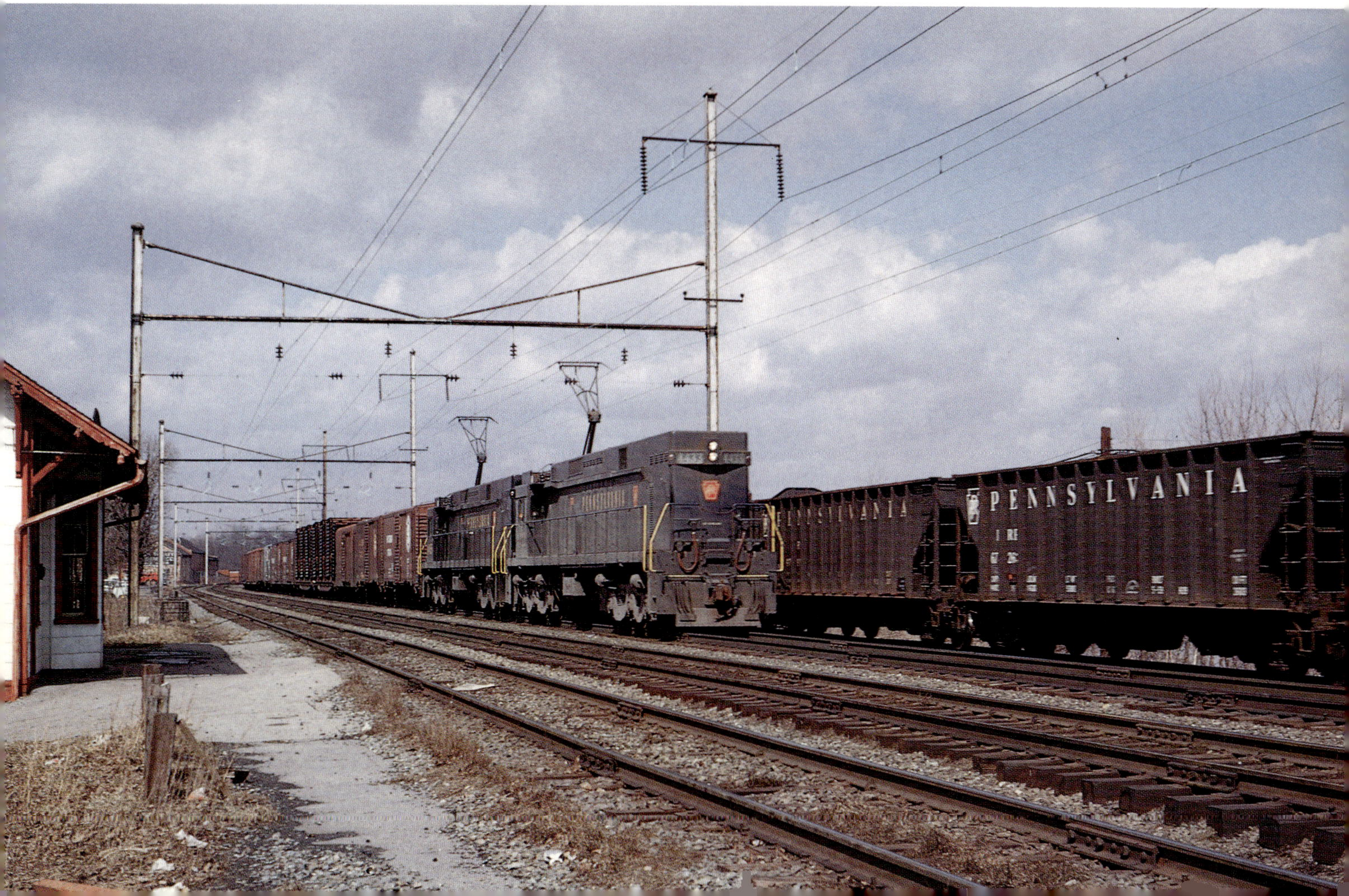
PENNSYLVANIA

Columbia & Port Deposit Branch

The C&PD Branch, as it was known, was a freight only line which basically connected Enola with Baltimore without the grades of the Northern Central line over which the passenger trains ran from Harrisburg to Baltimore. The NCRR was never electrified. In the 1950's and 1960's, the chief commodity hauled over the C&PD was coal with a couple of daily mixed freights each way between Enola and Pot Yard.

(Above) Beginning at COLA Tower in Columbia, Pa., where GG-1's 4825-4823 head south onto the branch October 1, 1961, the line followed the east bank of the Susquehanna River largely inaccessable by road and thus very difficult to photograph. Equally frustrating for photographers was the fact that the majority of the scheduled freights ran at night. Baltimore coal trains, being extras ran at any time, day or night.

(Opposite page, top) A trio of GG-1's 4842-4814-4818 are northbound through Washington Boro, Pa., one of the few towns on the line, April 8, 1961.

(Opposite page, bottom) This view could only be taken on a cloudy day because it looks south and shows a pair of E-44's coming west on the C&PD past Safe Harbor dam where much of the Pennsy electric power emanated from. The tracks above on the left were the beginning of the Atglen and Susquehanna Branch curving eastward towards Parkesburg and the mainline. The photo was taken in January 1968.

(Above and opposite page, top- Bill Volkmer, opposite page, bottom- Chet Fuhrman)

Harrisburg Yard

As a means of by-passing the congested Enola Yard and its hump, PRR made every effort to use the Harrisburg side of the river for any relay train such as coal and ore extras, trailer trains and the like. The photo *(opposite page, top)* of 4944 on February 16, 1960 and the accompanying photos depict Harrisburg Yard with its creosoted wooden catenary poles.

(Opposite page, bottom) As noted elsewhere in this book, the boss always wanted the P-5 modifieds to lead "wherever possible" and that is basically why the 4780 is seen leading the 4760 on January 29, 1961. However, an exception to this rule was made at the time the photo *(above)* was snapped. When the 4700 was dispatched as a lead unit on a South Amboy coal train enroute to a fantrip two days later out of New York, the box cab was leading so as to catch any bugs that may have crept into her during her mini-overhaul. It also gave the author/photographer an opportunity to shoot her before she got dirty!

(All- Bill Volkmer)

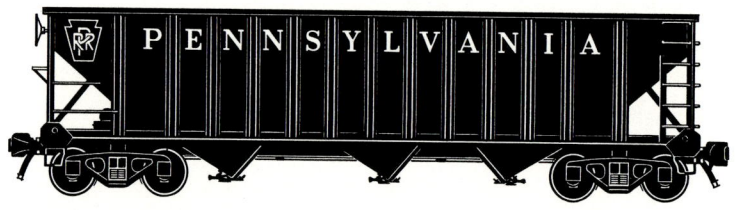

11523 254
B29

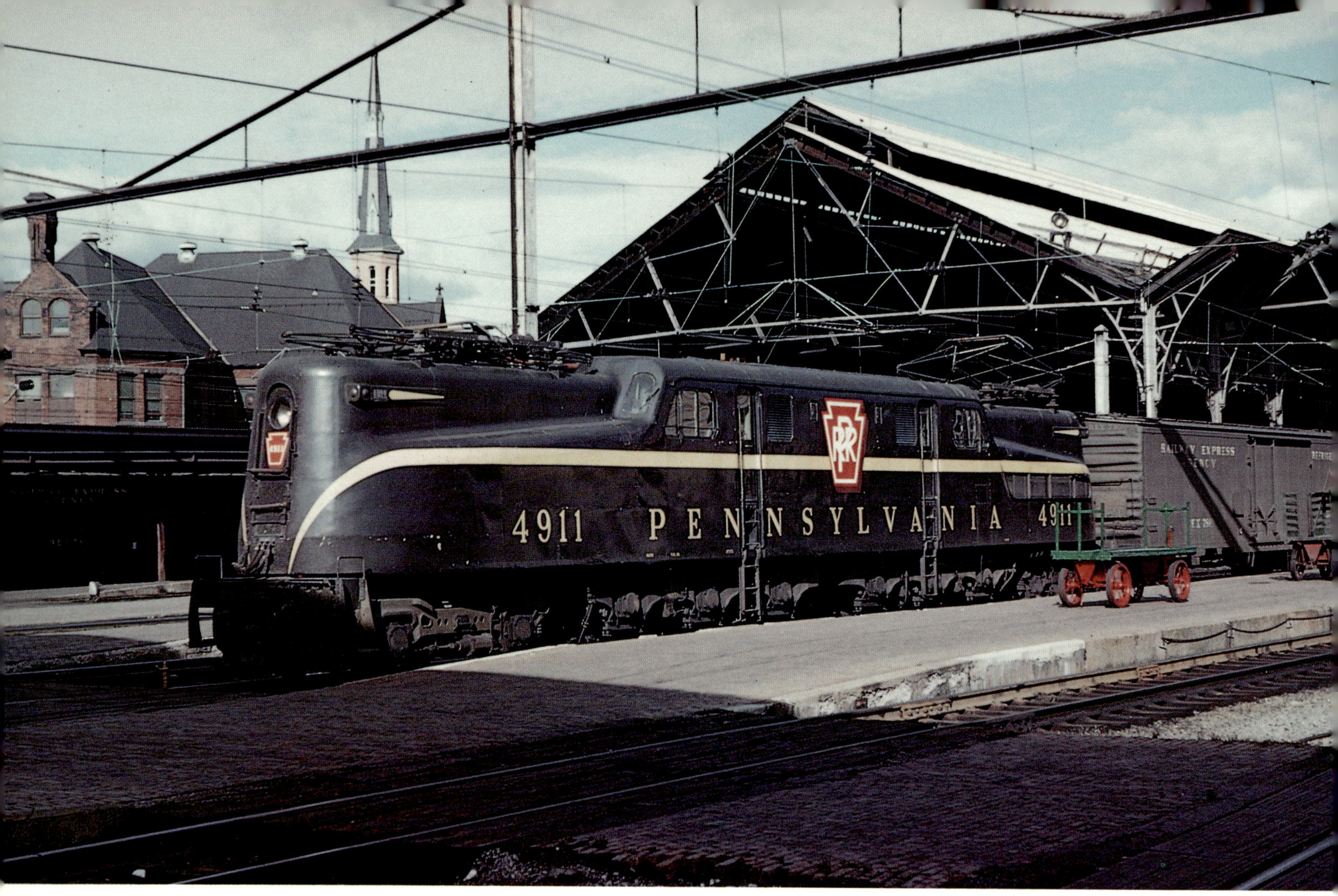
4911
PENNSYLVANIA
4911

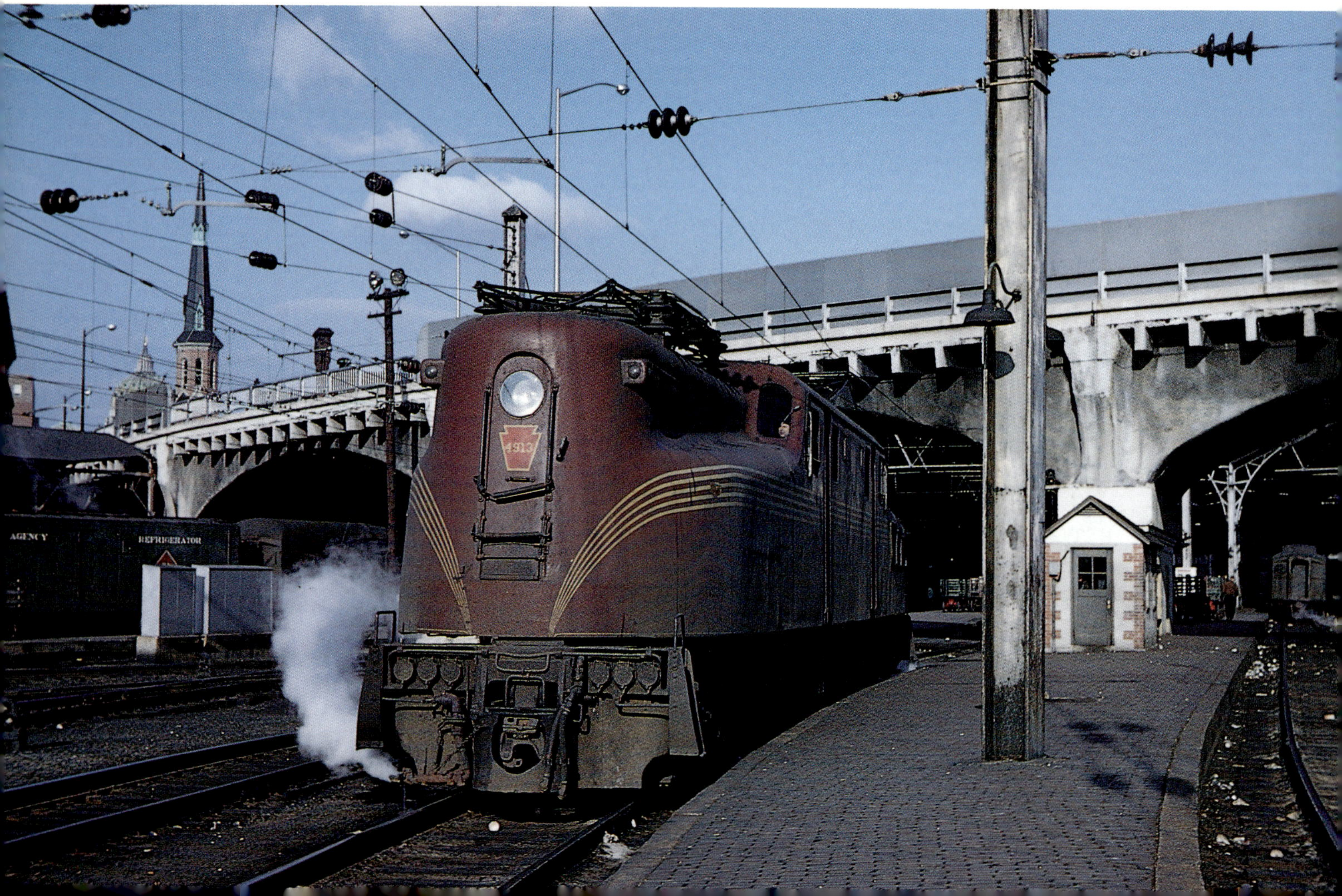
AGENCY
REFRIGERATOR
4913

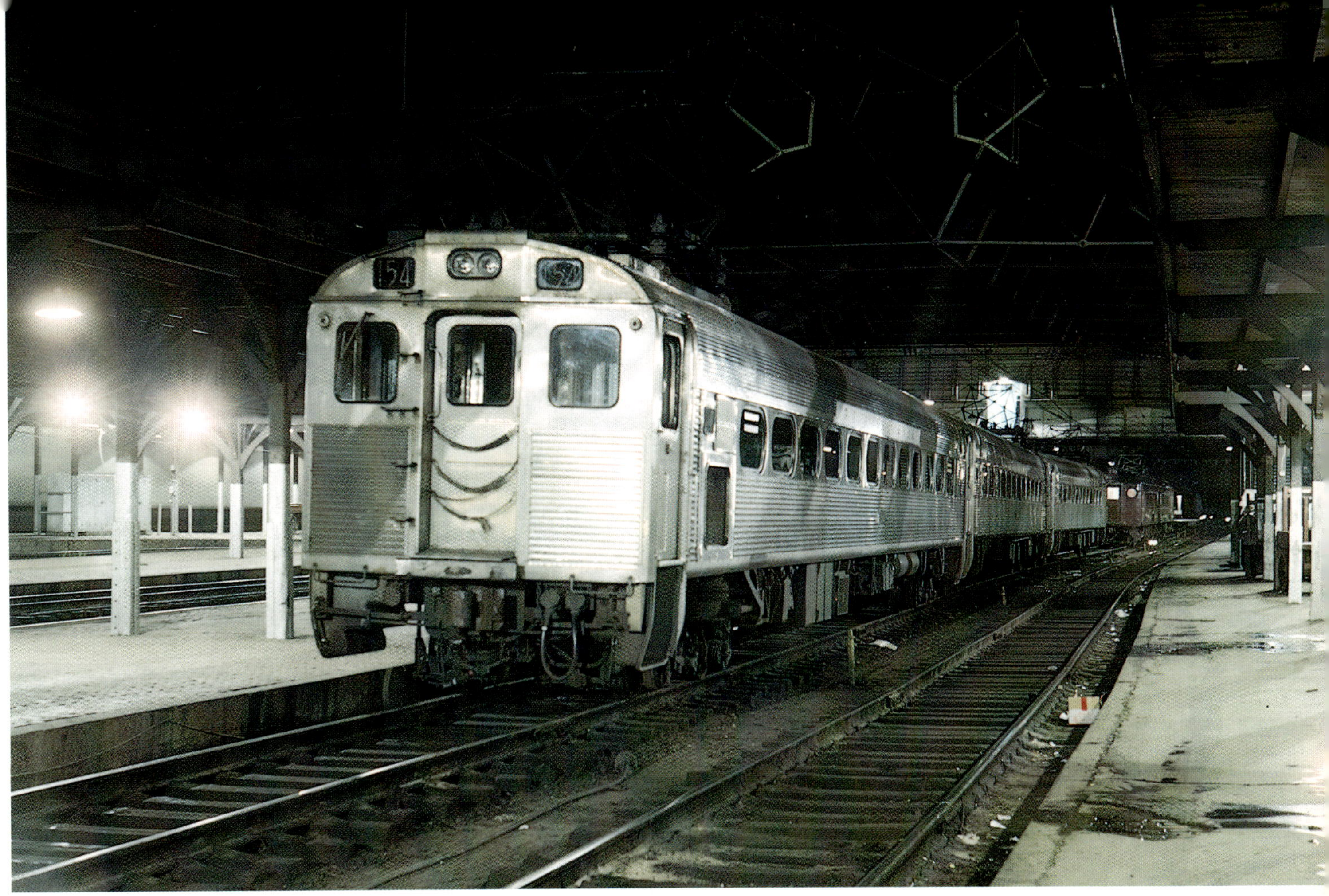

Harrisburg Passenger Station

Located 100 miles west of Philadelphia along the east bank of the Susquehanna River, Harrisburg, Pennsylvania's capital city, like all U.S. capitals, was the geographical center of the state's population when it was established as the state capital. It was also situated at the principal crossroads of all the main railroad lines in the state in the 1800's with lines radiating like spokes in a wheel, most of them PRR, in all directions. Because of the severe railroad congestion in downtown Harrisburg, Pennsy very early-on concentrated most freight operations in the area on what was known locally as the "West Shore," the Enola side of the river. The only freight traffic on the Harrisburg side of the river was that destined for interchange with the Reading and relay trains to and from the East. Virtually 100% of the north-south freight traffic passed through Enola and bypassed Harrisburg altogether.

Harrisburg passenger station was that point where the diesels came off and the electrics went on, and vice versa for the westbound limiteds. Because the inbound G's quickly departed for the servicing pits at McClay Street, the outbound G's were always easier to photograph. The 4911 is seen *(opposite page, top)* locking couplers with Train 50, the ADMIRAL from Chicago on the morning of October 15, 1961. Similarly *(opposite page, bottom)* the 4913, one of the last two pin-striped GG-1's, readies for departure on December 31, 1959.

(Above) Inside the trainshed in May 1965, three of the six original Budd Pioneer *Silverliner* cars await the morning run to Philadelphia.

(Opposite page, both- Bill Volkmer, this page, above- Bill Brennan)

#5687

Harrisburg

Station Switcher

During the daytime SW-1 diesels shunted coaches and head-end cars at Harrisburg, but at night the silent B-1's were used to add and detach the Washington sleepers for Trains 48-49, the GENERAL so that sleeping car patrons would not be disturbed. These three photos of B-1's are a study in contrasts. Using the Reading Station as a backdrop, *(opposite page, top)* we see the 5687 on March 25, 1960.

(Opposite page, bottom) The same little 5687, now with unusual red-painted windows sashes undoubtedly provided by an overzealous Enola painter, forms a Christmas Card scene using the Pennsylvania State Office Buildings as a backdrop, January 28, 1961.

And *(above)* three B-1's slept the day away on October 5, 1957 while a fourth B-1 switched the station. The year 1957 was the next to last big passenger train year for the east-west fleet and the last year for daytime use of the B-1's in Harrisburg.

(Opposite page, both- Bill Volkmer, above- Ken Douglas)

Eastbound

from Harrisburg

The mainline from Harrisburg to Lancaster was definitely not friendly turf for heavy eastbound freight trains because of the gradient east of Middletown. For this reason, those freights which bypassed Enola, being routed on the east side of the Susquehanna through Harrisburg turned south onto the Columbia Branch at Royalton and headed east via the "Low Grade Line" at Columbia and rejoined the main at Lancaster or at Parkesburg if the A&S Branch was the route of choice. Passenger trains always used the mainline through Mt. Joy to Lancaster and on east. Because most of the eastbound passenger action through this area was at night, about the only name trains which could be photographed on this line were the two depicted here.

(Above) Train 54, the PENNSYLVANIA LIMITED rolls east through Middletown, Pa., behind 4872 and *(left)* 4878 leads Train 50, the ADMIRAL past Leaman Place Jct., Pa. Both scenes were taken in July 1968 and both trains had their origination in Chicago.

(Both- Bill Brennan)

Leaman Place

In 1959, the name Strasburg Rail Road was merely a listing in the *Official Guide* under "S," nothing more, at least to the railfans of the day. Few fans had ever heard of it. The term "tourist railroad" then consisted of a couple of cog railroads and when a group of fans bought the Strasburg, I thought they were just plain crazy. Three years later, I found myself at Enola shop sending my one remaining boilermaker over to Strasburg once a month to do a real honest-to-gosh boilerwash on an operating steam locomotive! (See *Pennsy Diesel Years Volume 4,* page 14-15 for more on this).

The same lack of notoriety held true for Leaman Place Junction, near Paradise, Pa., where the connecting point between the nation's largest and smallest railroads lay. The two railroads, by the way, were identical in width! In the two photos GG-1's are seen wheeling fast freight past the junction.

(Above) GG-1 4875 is eastbound at Leaman Place in August 1962 and *(below)* 4829 leads a westbound at the same location, October 1, 1961.

(Above- Dave Sweetland, bottom- Bill Volkmer)

Thorndale

Helper Station

Helper stations, as we know them in today's diesel world, are places where extra power is added to the rear, or occasionally the center, of a train. This gives the train enough horsepower to make it up the grade which represents the ruling grade of the trip. Additional diesels are thus not required to go along for the ride just so they can be used on a short segment of the route.

Helpers in the electrified territory of the Pennsy took on a little different meaning. One of the principal differences between diesels, with their D.C. traction motors, and A.C. motored locomotives is that a diesel can literally start a train on a grade that it can't haul! At least not at any great speed. On the other hand, once an A.C. locomotive got rolling, it had little problem maintaining speed, even on grades.

The big problem with the P-5's and GG-1's, theirs being A.C. motors, was what to do if the train got stopped by a signal or hot box or other breakdown on a grade. Well, helper locos of the A.C. motor variety were the answer, of sorts, to management's prayers, because when they were idle at the helper station and the crew was off duty awaiting call, the locomotives burned no fuel (other than a trickle of juice to power the blower motors).

So when a train got stuck on a grade, or if the train was underpowered, the Thorndale helpers stood ready to go in all directions, east towards Paoli, west on the Atglen and Susquehanna Branch, west on the mainline, or east on the Trenton Cut-Off.

(Opposite page, top) FF-2's #3 and #6 rest at Thorndale on March 21, 1959. At this time these FF-2's had been on the PRR (from the Great Northern) about a year and a half and already they had expended almost half of their productive life on the Pennsy!

(Opposite page bottom) On a visit to Thorndale helper station on April 15, 1960, the author caught a rare occasion. Three fourths of the entire Westinghouse fleet was on one coal train getting a helper FF-2 for a run up the Trenton Cut-Off enroute to South Amboy. Units are 4995-4998-4996. The helper was FF-2 #6. An unidentified P-5 was also stationed there that day. Number 6 is ex-GN 5016, a G.E. product of March 1930.

(Above) A front view of the coal train shown on the opposite page. *(All- Bill Volkmer)*

Trenton Cut-Off

The Trenton Cut-Off was built from Morrisville, Pa., on the New York mainline, 46 miles westward to GLEN Interlocking on the Harrisburg mainline, skirting the Philadelphia metropolitan area on its northern perimeter. The line was built solely for freight traffic to avoid the congested Philadelphia terminal enroute to New York. To the author's knowledge, no regular passenger service was ever offered over the route. As late as 1936 there were no plans to electrify this freight-only branch or the Atglen and Susquehanna Branch (Low Grade Line) from Parkesburg west to Columbia and Harrisburg. To have continued steam operation on these lines would have caused the PRR to make huge investments in upgraded coaling stations, notably at Thorndale. An argument in favor of electrification was also made that the freight motors only ran 50 or so miles from Newark to Morrisville before they had to be removed in favor of steam. Thus it was that the Trenton Cut-Off became electrified in the big westward expansion of the late 1930's.

(Above) At Fort Washington, Pa., the Cut-Off crossed the Reading's Bethlehem Branch which was itself electrified as a lead to the Doylestown Line. It branched off the Bethlehem Branch at nearby Lansdale, Pa. The 4405-4407 were only a few months old when Al Holtz shot them westbound on the bridge. Date was May 1961. In years gone by there was also a branch connecting Fort Washington south to the Chestnut Hill Branch.

(Opposite page, top) One of the last P-5 assignments was trains EM-1/EM-2 known around Enola as the "Emmy." This turn ran east from Enola to Earnest Yard, near Norristown on the Cut-Off. Because the power layed over for almost 22 hours before returning to Enola, the oldest locos on the roster were always assigned. Here the 4711-4742 make one of the last-ever P-5 visits to Earnest Yard in January 1964.

(Opposite page, below) Al Holtz also caught the 4772-4727 eastbound on the Cut-Off in the 1961 time frame. The 4772 was totalled in a wreck in August 1962 at Parkesburg when it rear-ended a stopped coal train.

(Opposite page, top- K. Allen Keller, opposite page, bottom and above- Al Holtz)

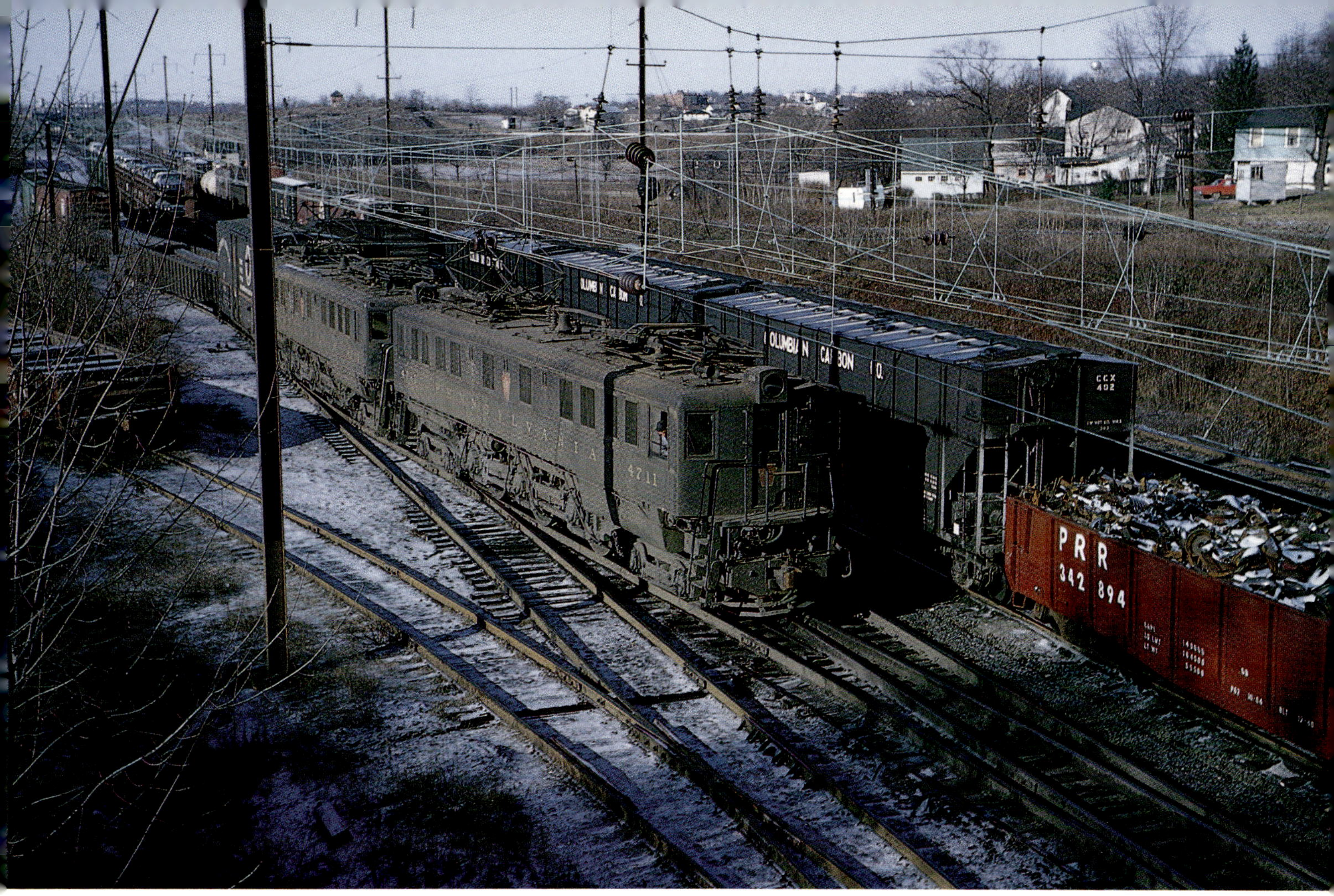
PENNSYLVANIA
4711
CCX
402
PRR
342 894

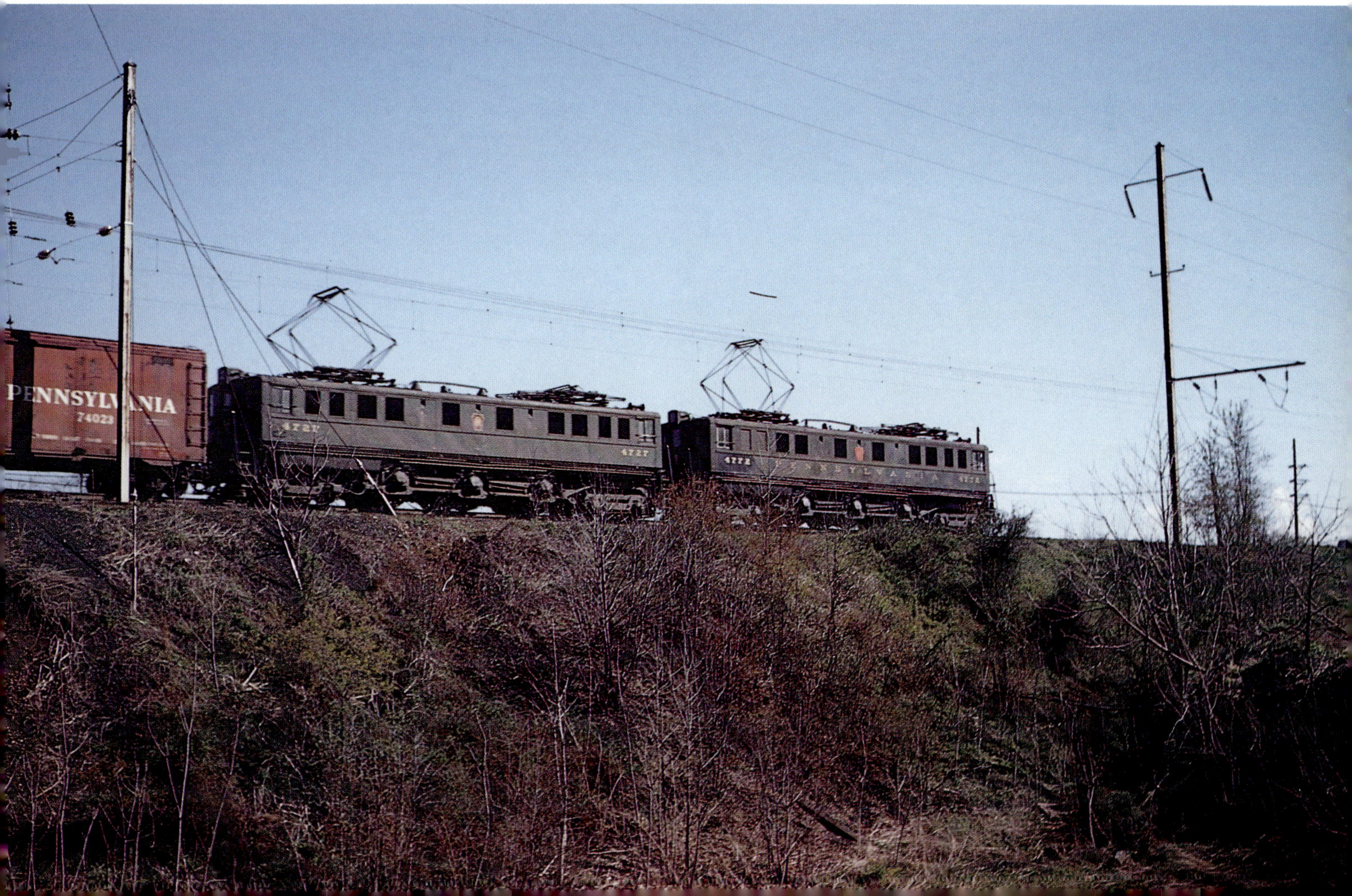
PENNSYLVANIA
74023
4727

MU's to Harrisburg

In 1961, Pennsy was under pressure from the State Capital to put on a commuter train with state subsidy to cater to Harrisburg and Lancaster area commuters who worked in Philadelphia. So a single MP-54 E-6, among the first to be equipped with automatic train stop for single car operation, was assigned to Trains 602/605 leaving Harrisburg at 5:35 AM and returning at 7:25 PM. The car laid over each night at Harrisburg. By 1965 when Bill Brennan paid a night visit to Harrisburg station, patronage had increased to where a second train had been added and two and three car trains were being used.

(Above) A two car set of MP-54's lays over in the station in May 1965. *(Left)* The 1963 order of Budd *Silverliner II* single unit cars was bought in part by the PSIC to equip the Harrisburg operation. Here car 203 is just out of the builder's plant in May 1963 prior to delivery to PRR.

(Above- Bill Brennan, left- R.S. Short)

Paoli Shop

Every one of the 400-plus fleet of MU cars operated on the Pennsy paid a visit once per month to the Paoli Car Shop for preventative maintenance. This required a daily "deadhead" train each way from New York usually with a stop at New Brunswick and Trenton. Once per month the deadhead train would make a stop at Princeton Junction to pick up the "dinky" for a trip to Paoli. The "dinky" was a pair of MP-54's which shuttled back and forth between Princeton Junction and Princeton and as such were more or less isolated from the rest of the system.

(Above) On April 12, 1960 one of the two MBM-62E express motors, number 5288, was receiving its monthly checkup at the shop. It and the 5290 hauled newspapers from the *Big Apple* to Trenton.

(Below) A very rare variety of MU car is illustrated below in back of Paoli Shop. Car 517 is one of six motorized trailers. These MP-54 E-1's were numbered 497, 498, 510, 511 and 517, 518. All were scrapped by June 1960. The 517 was captured on film in March 1959.
(Above- Bill Volkmer, below- R. S. Short)

Route of the Paoli Local

In 1912, two years after Penn Station, New York opened for business (as a D.C. operation) the PRR announced intentions to electrify using A.C. from Broad Street, Philadelphia up the hill to Paoli, some 20 miles west. The MU cars which began operating over the route in September 1915, PRR's very first A.C. operation, not only immensely increased the capacity of Broad Street Station because of the double-ended nature of the MU cars, but the sure-footed electric cars shaved fully ten minutes off the 59 minute run while making all stops. It was the huge success of this operation that caused the railroad to become enthusiastic about electrifying the New York to Philadelphia and the various other Philly suburban routes immediately following World War I. The grade up to Paoli always required helpers on heavy ore trains from South Philadelphia, usually added at ZOO Tower out on the High Line. The downhill run on the other hand required special dexterity on the part of the engineman with the brake valve when the power was GG-1's or P-5a's. This was especially true of the P-5a's because in addition to their lack of dynamic brake, the hot brake shoes were riding up against 72" diameter tires just like the old passenger steam locomotives, which never hauled coal trains! If the engineman rode the brakes to the point of heating up the tires, the tires would expand, and frequently loosen the tires on the wheels! On these pages are scenes along the route from ZOO Tower to Paoli.

(Opposite page, top) BRYN MAWR Interlocking served as a sort of plateau in the climb between ZOO and Paoli. An ore train grinds uphill through the interlocking April 8, 1960. The open switch on the siding in the foreground is for the wire inspection car which was in the hole to let traffic pass. The cameraman was on board that car.

(Opposite page, bottom) Another ore train heads west through Narberth with E-2b's 4939-4943 and *(above)* 4722-4724-4735 haul yet another ore train past Merion, near the beginning of the ascent. Date is May 5, 1962.

(All- Bill Volkmer)

Merion

Home of V.P.'s

If there is any one town along the "Main Line" west of Philadelphia which typifies that term (i.e. home to Philadelphia's upper class management types), then Merion would be a prime candidate. It is the first stop west of the Philly city limits and a great place to take photos during the late spring when the sun comes up at 5:15 A.M. and the azalias and dogwoods are in bloom. Indeed, even as these words are written, one may wander into the clubhouse of the exclusive Merion Golf Club nearby and probably overhear a couple of retired PRR moguls (the human type, of course) discussing "the way it used to be."

In the three accompanying photos one may note that Merion is also the bottom of the long grade to Paoli. Since the GG-1's, P-5's and the MP-54 MU cars were not equipped with dynamic brakes, brake shoe smoke tended to permeate the station area when eastbound trains were present.

(Opposite page, top) P-5's 4716-4732 are rolling east with a manifest freight before 7 A.M., May 6, 1962.

(Opposite page, bottom) The previous morning, the author caught 4722-4724-4735 westbound with an ore drag. The two unit Baldwin centercab helper appeared in a photo published in *Pennsy Diesel Years Volume I,* page 32.

(Above) An early morning commuter train screeches to a halt at the Merion Station, May 6, 1962.

(All- Bill Volkmer)

ZOO Interlocking
and the "High Line"

Probably the single most significant interlocking tower in the entire United States, let alone on the PRR, was ZOO Tower in northwest Philadelphia. This tower governed moves at the junction of the PRR east-west mainline with the New York to Washington mainline. Trains could be routed to Suburban Station via Powellton Ave., to 30th Street Station and Penn Coach Yard, and onto the High Line to ARSENAL Tower. Trains on the High Line could bypass the entire 30th Street complex on their way to South Philadelphia and Washington. Because of the complexity of the number of lines which interchanged with one another, the interlocking had an unusual array of flyovers, duck-unders, subways, viaducts and so forth. The interlocking was so complex that it was possible for trains to pass through on some routes unseen by the operator in the tower for more than an instant. Modern day freeway interchange designers have nothing on the architects of ZOO Interlocking who did their work prior to the turn of the century long before freeway interchanges were invented.

(Above) A Food Fair grocery *Truc Train* hauled by lead GG-1 4824 passes the tower northbound off the High Line at the same time as Train #33, the ST. LOUISIAN ducks into what was called "the New York and Pittsburgh Subway" on its way to Paoli and St. Louis. The date is May 6, 1962.

(Opposite page, top) GG-1 4829 still in the five-stripe scheme leads a southbound freight past the Philadelphia Zoo, from where the interlocking received its name, on July 17, 1958. The single stripe paint scheme was inaugurated in 1956 and by this date, the majority of the G's had already been repainted.

(Opposite page, bottom) Another angle on the tower shows Train 30, the SPIRIT OF ST. LOUIS in June 1969 with the 4904 emerging from the subway underpass. Westbound tracks from 30th Street to Paoli passed above.

(Opposite page, top and this page- Bill Volkmer, opposite page, bottom- Bill Brennan)

Sunnyside Yard

Sunnyside Yard in its heyday was unquestionably the world's largest and busiest passenger coach yard. Situated in New York's Queens borough, it was the point of countless daily departures for all points near and far, to Miami, Chicago, St. Louis and South Amboy. Several tracks in the yard were equipped with third rail even in later years for use by the Long Island R.R., but in modern times these tracks were given over to storage of mail cars. There was also a balloon track for turning the trains and washing them simultaneously, using in most cases the inbound G for power.

Here is a sampling of the sights to be seen around Sunnyside.

(Above) The 4933 leaves the yard with the 1956 Budd-built tubular train known as the *Keystone*. This was in 1967. In earlier years this would have been a rare find as that train was based at Philadelphia and always turned at Penn Station in New York for its return trip to Washington. Each day it made 2 1/2 round trips to Washington and was serviced at Penn Coach Yard, Philadelphia during the night. *(Below)* GG-1 4909, probably assigned to the South Amboy shuttle service sleeps the night out at Sunnyside enginehouse.

(Above) B-1 3913 switches mail storage cars in front of the commissary in November 1966.

(Below) GG-1 4885 heads westbound into the East River tunnel at Hunter's Point in December 1967. *(Both pages- Bill Brennan)*

Masterpiece Polish

Masterpiece Polish was the trade name for a creamy commodity that came in five gallon buckets and select Enginehouse Foremen around the railroad used the stuff on their locally assigned switchers as a sort of symbol of local pride, a very rare thing on the deficit ridden, deferred maintenance Pennsy of the last few years. In those days about the only time the GG-1's were hand polished was when they were assigned to the Army-Navy Game Specials. Road diesels used in later years to the game also received this treatment.

When VIP's were going to ride, the polish at Sunnyside normally reserved for the B-1's and L-6's, was hauled out for the GG-1's assigned to the VIP special. One such occasion was March 16, 1960 *(left)* when the 4916 was polished up for, of all things, a nighttime run west on the BROADWAY LIMITED hauling a group of insurance company executives. Incidentally, the 4916 and the 4907 were the only two GG-1's to ever receive the Tuscan Red paint scheme incorporating the single yellow stripe.

(Below) B-1 3919 at the same location shows off its polish job the previous day. The 4841 on the left is shiny owing to its recent repaint at Wilmington Shop.

(Both- Bill Volkmer)

New York,
New York

Penn Station, New York was for all practical purposes the birthplace of the Pennsy electrification. For years all the railroads which began and ended on the New Jersey side of the Hudson had been scheming how to build a bridge across the river and put a terminal in New York. There were a number of obstacles to be overcome in order to build a bridge over the Hudson. The city was already choking in steam locomotive smoke, there was precious little real estate (above ground) even in the Gay 90's, which could be devoted to trains, coaling stations, water plugs and the like, and to add insult to injury, they would have to tunnel through Bergen Hill on the approach to the bridge. Furthermore, a bridge would require an elevated station to be built in midtown Manhattan. In 1901, Pennsy bit the bullet and laid plans to dig under the Hudson, under New York City and under the East River.

The result after nine years of construction was an electrified, albeit 3rd rail D.C.-powered station and yard of monumental proportions. The station ediface, opened in 1910 was located between 31st and 33rd Streets, 7th and 8th Avenues. It was one of the most architectually graceful buildings in all of New York City and when it was torn down in 1966, (replaced by the present day Pennsylvania Towers and Madison Square Garden) it was sorely missed by students of architecture.

(Above) The Ninth Avenue facade of the Post Office which was directly across Eighth Avenue from the station shines in the late afternoon sun as the 4917 heads west with but one of hundreds of daily departures in post-PRR, April 1968. Penn Station as a building followed the great railroad into oblivion.

(Bill Brennan)

Under the Hudson

Few railfans ever had the chance to walk from Penn Station to New Jersey with the tunnel inspector and a camera in tow. The author did just that on the morning of March 2, 1960 while assigned to New York as a Junior Engineer. In those days, the GG-1's never ventured onto New Haven rails, but the New Haven passenger locos did tread on Pennsy rails, coming into Penn Station over the Hell Gate Bridge and past Sunnyside through the East River Tunnel.

So this photographer was quite elated to be able to shoot a GG-1, 4896 *(opposite page, top)* sunning itself beside New Haven EP-4 #360 just west of the station throat. The 360 was originally the 0366 when built in May 1938. The 4896 was 11 months younger, having been outshopped from Altoona in April 1940. At the time this photo was made the photographer mused that if the PRR were to for some reason buy the 360 series locos from the NH, they would logically be called "GG-2's." I wasn't quite sure I could handle that.

(Opposite page, bottom) The 4884 with a heavyweight sleeper in tow emerges from the tunnel into the daylight.

(Above) Across the river in New Jersey, the tunnel emerged at North Bergen where this photo of GG-1 #4902 popping out with the SILVER METEOR was taken in April 1968.

(Opposite page, both- Bill Volkmer, this page, above- Bill Brennan)

Bergen Hill

Leaving the Bergen Hill Tunnel after crossing under the Hudson River, passenger trains exiting New York City travelled across the marshlands known as the Jersey Meadows. The photo *(above)* shows a G in 1967 passing the remains of the recently torn-down Penn Station which had been dumped as landfill in the Jersey Meadows the previous December.

(Right) A clocker crossing the Jersey meadows at a good clip towards Newark in October 1968. After WWII, the third rail which once existed here was removed and ended at the Bergen Hill Tunnel west portal.

(Above- Al Roberts, right- Bob Yanosey)

Manhattan Transfer

Manhattan Transfer was the point south of the Hudson tunnels where the D.C. third rail electrification ended in the 1910 installation project, and steam locomotives were added to the southbound passenger trains. The advent of A.C. electrification ended the need for both steam locomotive substitution and the need for any semblance of a PRR station at this location.

After the A.C. overhead catenary went up over the New York Division in the early 1930's, Manhattan Transfer more or less ceased to exist and thereafter became the H&M's Harrison station. On December 31, 1960, the 4889 hauls the Lehigh Valley Railroad's Toronto to New York MAPLE LEAF on one of that train's final runs. These coaches inspired the *American Flyer* brand toy passenger cars, or was it vice versa?
(Both- Ken Douglas)

Exchange Place

Those who have read *Pennsy Diesel Years Volume II* noted the cover photo showing a Baldwin Shark poised to take a train to South Amboy and points south. Usually on the adjacent tracks sat hoards of MU cars also awaiting the arrival of the "5 o'clock sailors," as the trans-Hudson passengers were sometimes referred to, for a run as far south as the wires went, South Amboy. *(Above)* The 758 which leads this train is an MP-54 E-2, one of a group of 40 MU's built in 1927 by the Standard Steel Car Co., of Hammond, Indiana. Of all the MP-54 cars owned by the Pennsy, 740-819 were the only group (80 cars) that were built new as MU cars. All the rest were converted at one time or another from steam coaches. The date of this photo, February 27, 1960 was long after the Pennsy ferries had ceased operation and all those "sailors" now came by H&M train to the below ground station.

The 667 *(below)* rounding the curve at Elizabeth is an MP-54 E-5, one of a group of 49 cars modernized with new seats, lighting and most importantly, trucks/traction motors in 1950-51 at Wilmington Shops using Westinghouse traction equipment. Date for this photo is April 12, 1968.

(Above- Bill Volkmer, below- Bob Yanosey collection).

Journal Square

The P&H Branch

Journal Square was a junction point on the Pennsy for the lines to Exchange Place and Harsimus Cove Yard. It was a good place to watch PRR action around WALDO Tower which governed moves through the area. Light engine-and-cab moves were also to be seen as the crews ferried to and from Meadows Enginehouse which served all the various north Jersey yards: Harsimus, Kearny, Greenville, and Waverly.

(Above) A set of E-44's lead by the 4406 trundles over the Hackensack River Bridge enroute to Harsimus Cove Yard to pick up a merchandise freight in October 1966. (Below) Freight G 4839 skirts the rock outcroppings in the shadow of the H&M Journal Square station in January 1965. (Both- Bill Brennan)

Hudson & Manhattan R.R.

The H&M tunnel under the Hudson River at Exchange Place was actually dug and opened prior to the start of the digging of the Pennsy Bergen Hill Tunnel to Penn Station. It was begun in 1901 and the Pennsy found itself bankrolling the financially ailing H&M much of its checkered career until the Port Authority Trans Hudson took it over in 1962. The Pennsy did not wish to get into the subway business, but because the H&M gave PRR customers fast access to lower Manhattan, without enduring the rigors of the ferryboats which were very cold in the winter, it felt compelled to keep H&M going.

In the late 1950's the line was still using its original fleet of cars and PRR again came to the rescue by purchasing new cars from the St. Louis Car Co. *(Above)* An H&M car classed as MP-51 sits at Newark Station in October 1963 when brand new. *(Below)* Even newer yet is a pair of H&M cars being delivered from the builder through Altoona yard, September 13, 1958.
(Above- R. S. Short, below- Bill Volkmer)

Elmora Tower

About a half mile south of Elizabeth, New Jersey stood ELMORA Tower. This represented the interlocking where the four tracks headed south from Newark became six, preparatory to the PA&W-South Amboy line taking off at UNION Interlocking in Rahway. Tracks 1, 2, 3 & 4 had tracks A & B added. A & B tracks were for the South Amboy and other trains that made stops at Linden while the regular tracks carried the through traffic. The photos here catch the tower in two different moods. *(Above)* ELMORA plays host to a southbound mail and express train headed by GG-1 4882 in October 1962 while *(below)* E-44 4409 leads a southbound TTX train in April 1965.

(Both- Bob Malinoski)

Elizabeth Curve

One of the New York Region's most photogenic curves was the sweeping S curve at Elizabeth, N.J. Publicists for the Pennsy quickly learned how to best emphasize the graceful lines of the GG-1 in both morning and afternoon light, on both sides of the track. Ken Douglas caught one of Pennsy's many and varied types of passenger consists, the SILVER METEOR from Miami *(above)* with pin-striped GG-1 4934 leading on August 11, 1951.

(Ken Douglas)

Jamesburg Branch

South Amboy's other entrance

One of Pennsy's least photographed electrified lines was the Jamesburg Branch. This was because few trains ever used it and all those that did, ran as extras. Chief commodity was coal eastbound and empty hoppers westbound, as well as general merchandise traffic for PRR's Brown's Yard in Old Bridge and the partially owned Raritan River Railroad in South Amboy. It should be remembered that in the 60's there were no radios and scanners to help the poor railfans out when chasing trains to photograph! So when the NRHS commemorated the 25th Anniversary of the first GG-1 being outshopped on May 17, 1959, a fantrip was run using the freshly repainted 4800 (with suitable coomemorative plaque affixed) over the branch. The photo *(above)* shows the train at a photo stop in Jamesburg, New Jersey.

(Ken Douglas, Volkmer collection)

Greenville Branch

Situated approximately 3 1/2 miles south of Newark Station was LANE Tower, junction point for the 5 1/2 mile Greenville Branch serving the car floats at Greenville Yard and Waverly Yard. Greenville Yard was the point of embarkation for the Pennsy navy to transfer freight to points on the east side of New York Harbor chiefly to the Bay Ridge Terminal in Brooklyn where a connection with the New Haven R.R. was made via the Long Island R.R. The heavy passenger train traffic through Penn Station coupled with the extremely tight clearances in the tunnel and station precluded the routing of freight trains across the Hell Gate Bridge enroute to New England from the Pennsy. It is interesting to note that a 1962 timetable shows 12 scheduled freights per day terminating (eastbound) at Greenville, but only 7 trains being dispatched (westbound). This reflects the role of New York as a consumer of loads and generator of empty cars and also tends to account for part of the seemingly endless parade of light engine moves over the electrified region.

(Top) The 4840 heads out of Greenville in November 1965 with a mixed bag of freight typical of that which came off the floats. *(Bottom)* At the junction with the mainline at LANE, the 4808 gains

momentum on its trek south towards Philadelphia in the summer of 1968.

(Opposite page) Two views of a rare event on the Greenville Branch, a *Rotary Club Special* crossing the Newark Bay bridge enroute to interchange with the New York Central for a trip up the West Shore of the Hudson. #4928 is in charge this day. The 4928 is an example of a GG-1 with the centrifugal snow filters applied as evidenced by the large air intakes in the hood under the pantograph.

(Opposite page, both and this page, top- Bill Brennan, this page, bottom- Bob Yanosey)

Princeton Junction

Princeton Junction, always a railfan favorite place, was the station/junction where the two car MU train to Princeton University connected with trains from New York and Trenton. The shuttle was affectionately known as the "dinky" and the two cars assigned there were captive for a month at a time. Because the line had only one curve, this caused the wheels to wear out on one side of the train at an inordinately high rate. The three mile trip to Princeton took all of six minutes and ran at intervals from twenty to forty minutes depending upon the time of day and frequency of connections. In today's airline parlance, most all runs were carded as "non-stop." In this photo montage we accompany photographers Bill Brennan and Bob Yanosey on a December 1967 outing to the junction.

(Above) GG-1 4903 is southbound with the SILVER METEOR roaring past the shadows of Brennan and Yanosey on the platform.

(Opposite page, top) Earlier that day, a two unit E-44 team with 4440 leading, passes Nassau Tower visible in the distance at right.

(Opposite page, bottom) The dinky lays over on the Princeton Branch that afternoon waiting for connecting passengers.

(All- Bill Brennan and Bob Yanosey)

Trenton

New Jersey's capital city was an interesting spot for the train watchers. The continual passing of north-south passenger limiteds was augmented by not one, but two doodlebug runs, one to Camden, the other to Red Bank (see *PDY II*, page 17, *PDY IV*, pages 28-29). Interesting MU consists with R.P.O's and MU combines, diesel freighters in and out of Morrisville and the Bel-Del Branch rounded things out.

(Above) An inbound MU train lead by a bride-and-groom trailer-combine duo with the combine not leading is about as unusual as one can get. The date is October 3, 1959.

(Opposite page, top) A photographer who was careful with the stealthiness of the GG-1's, braved the track center in April 1968 to catch a good shot of 4872 southbound. Doing this shows us how the Trenton station was laid out.

(Opposite page, bottom) The 4927 leads Train 49, the GENERAL, across the Delaware River Bridge just south of Trenton station in July 1963. The first car back of the loco is a crew dormitory car. Train is enroute to Chicago overnight.

(Above- Bill Volkmer, opposite page, top- Bill Brennan, Yanosey collection, opposite page, bottom- Bill Brennan)

Chestnut Hill Branch

The heaviest patronized commuter run on the PRR was the 6.6 mile Chestnut Hill Branch local which operated entirely within the City limits of Philadelphia. In fact, the end of the line coincided with the end of the Philadelphia Transportation Company's Route 23 Germantown streetcar line! Despite the fact that the line operated within the confines of the big city, it was actually quite scenic on its outer end, crossing the high trestle *(left)* at St. Martins in this August 1967 photo, and *(below)* stopping at the second inbound stop, Highland in June 1963. Car 767 leads this train. One little remarked branchline in Pennsy electric territory was the Fort Washington Branch or Cut-Off as it was sometimes called. This line branched off the Chestnut Hill Branch at Allen Lane and operated two round trips per day to Whitemarsh, 6.5 miles away. It was de-electrified in the late 1940's after MU service was discontinued.
(Both- Rich Short)

Passenger Service Improvement Corporation

When the Pennsy commuter service in Philadelphia found itself in dire financial straits in the late 1950's, the City of Philadelphia decided to do something about the badly deteriorating service. Effective with the October 26, 1958 timetable, it set up what was to become the Passenger Service Improvement Corporation to subsidize the operation while lowering fares and coordinating transfers on the Chestnut Hill line with the streetcars and bus lines it intersected. They also agreed to finance some new MU cars to be operated on the PSIC backed lines as a means of boosting patronage. In 1963 the first 38 *Silverliner* cars were bought from the Budd Company in order to keep the tax money local (along with 17 similar cars for the Reading). A later group of cars surprisingly was bought from rival St. Louis Car Company in 1968. These were delivered with Keystones even though the PC merger had already taken place. When SEPTA (Southeastern Pennsylvania Transportation Authority) came into being in 1965, further improvement and integration with the city transit division was accomplished.

(Above) This photo illustrates the origin of the name *Silverliner*. It was taken at Jenkintown while the brand new unit was being tested by Budd on the Reading Company's tracks in July 1963.

(R. S. Short)

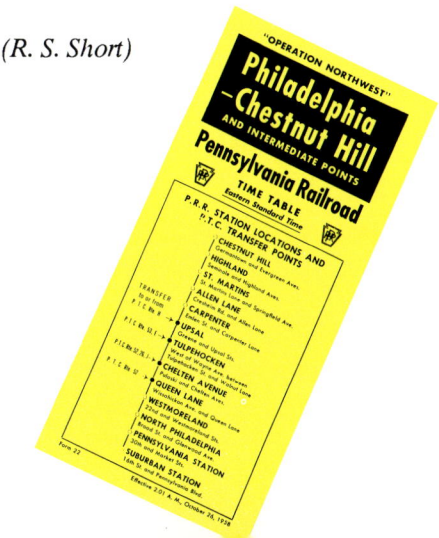

Cab Signals and Train Stop

Many fans wondered why the red MU cars had a single red stripe in the end windows and, in later years, a double red stripe. The reason for the stripes was to identify which MU's had cab signal equipment (single stripe) and which were equipped also with automatic train stop (double red stripe). The train stop was applied to the cars when the State of New Jersey mandated that all trains operating in that state be equipped with the train stop feature resulting from a CNJ commuter train running off the Elizabethport-Newark Bay Drawbridge in 1958. The photo *(above)* shows MU car 677, one of the MP-54 E-5 (1950 modernized) class emerging from Bergen Hill Tunnel in February 1968 with its double stripe visible.

Of course, all the GG-1 locos had train control equipment on board, long before the drawbridge incident but for some strange reason, only this one P-5a, the 4727 was so equipped. Because it was, #4727 was always supposed to lead, even if there was a modified in the consist. On April 29, 1961 when the 4727 was snapped going past North Philadelphia *(left)* it had no trouble leading owing to the fact that this was a solo flight for her.

(Above- Bill Brennan, below- Bill Volkmer)

Fairmount Park

Philadelphia

Fairmount Park was one of the nation's largest city parks and was notable for having its own trolley line (abandoned in 1946), a zoo, horse trails, exposition, police force on horseback, rowing regattas, and the like, but to railfans it also was a great place to photograph the Pennsy action through Philly. About midway between ZOO Interlocking and Margie Yard very early morning light in early May allowed one to catch the eastbound BROADWAY LIMITED which much of the year made its entire New York-Chicago-New York run under cover of darkness. Come with us then on the Saturday morning of May 5, 1962 to the park and spend one hour with us while we watch part of the morning "rush." At exactly 8 o'clock on #1 track *(above)* the BROADWAY, Train #28 with 4919 in the lead slams past, right on time as it usually was. *(Below)* Ten minutes later on #2 track a Greenville-bound freight, MD-6 (Maryland Division) headed up by 4403-4404-4407 passes. The phenomenon of general merchandise freight as seen here is almost as extinct today as the E-44's that hauled it! When you count the late running GENERAL and some MU's also photographed, it was all in all, not too bad for an hour trackside! *(Both- Bill Volkmer)*

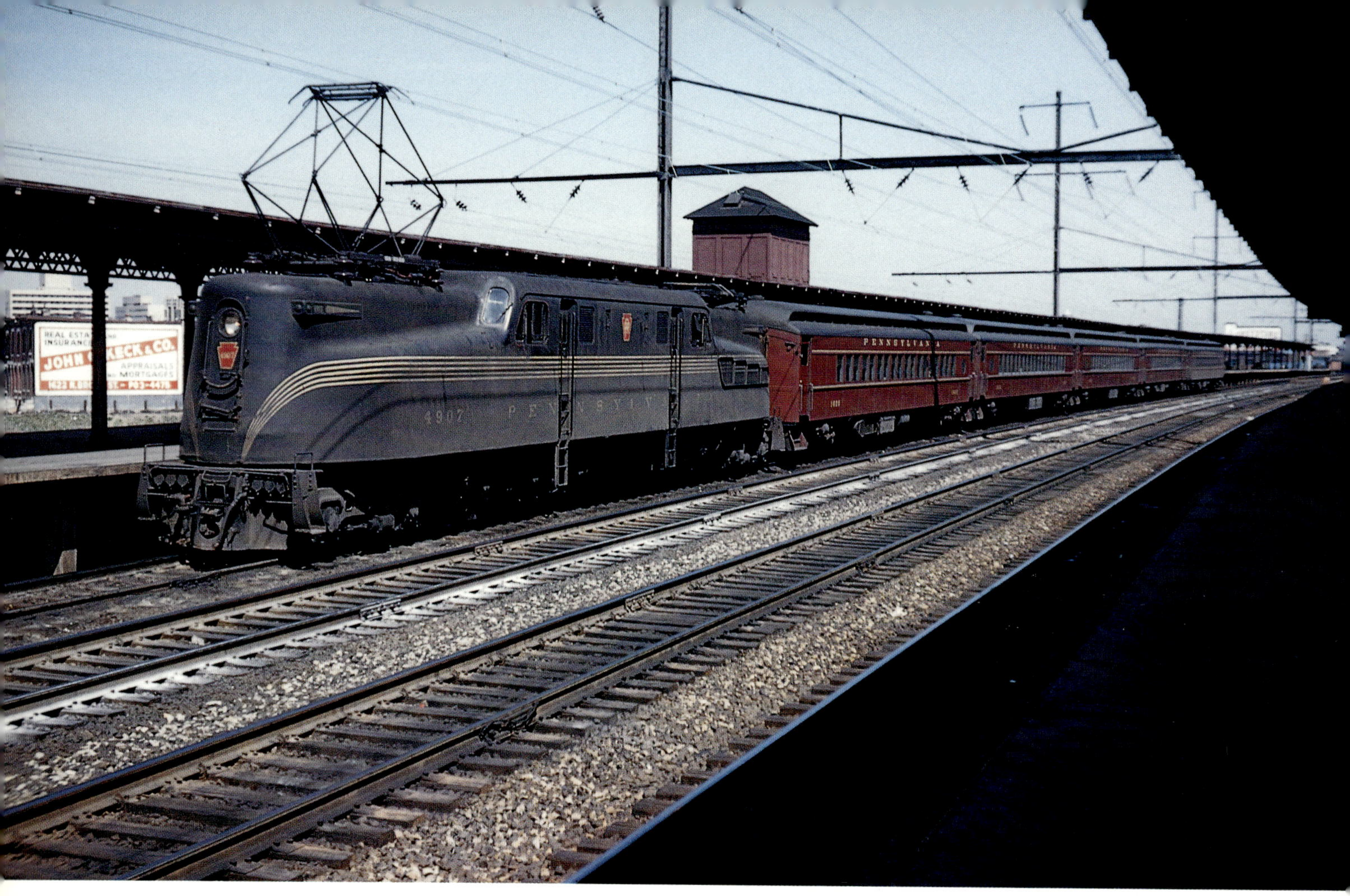
APPRAISALS
MORTGAGES
4907
PENNSYLVANIA

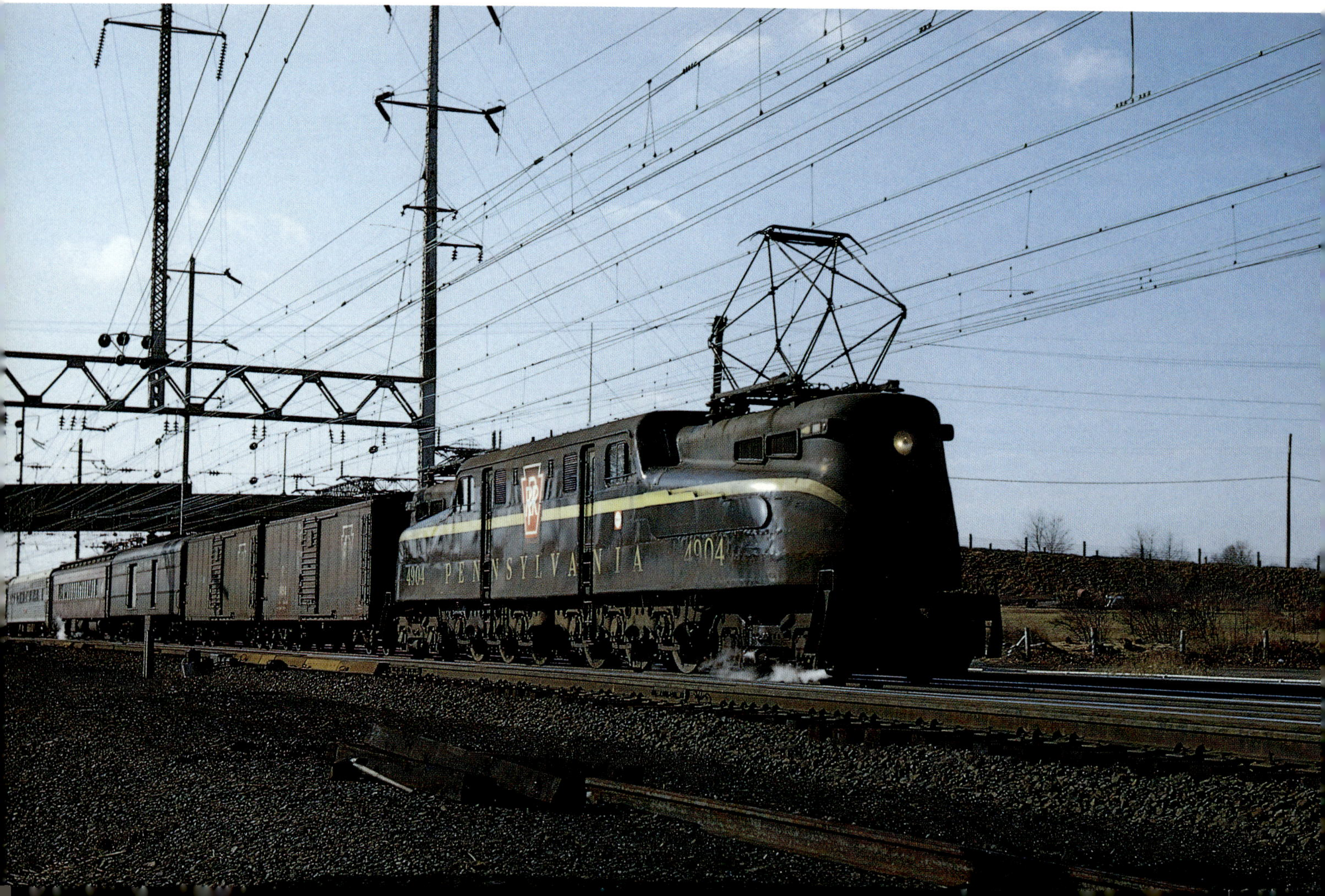
4904 PENNSYLVANIA 4904

The Clockers

Every hour, on the hour, during the daylight hours a GG-1, from six to ten P-70 fbr's, and usually one or two parlor cars would leave Philadelphia for New York. Naturally a similar service was operated in the reverse direction but not necessarily on the hour. It was sort of the Pennsy version of the modern day air shuttles. New York fans used to refer to the Hudson and Manhattan as "the little red subway" and the PRR as "the big red subway." Some of the clockers operated through to Washington in later years including the unique "Keystone" train. Weekdays, the *Keystone* equipment was tacked on the rear of a southern consist because the HEP unit needed to accompany the cars. However, on Saturdays, the *Keystone* operated as a separate consist. On Saturday July 22, 1961 the author poised with camera *(below)* at North Philadelphia to catch a "pure" *Keystone* consist directly behind the engine. Murphy's Law came into play and there was a P-70 coach full of school teachers added ahead of the tubular equipment to mar the photo, but they can't say I didn't try! In the days of steel pantograph shoes, the fireman got out of the locomotive cab at every station stop to look up and inspect the shoe for the presence of holes.

(Opposite page, top) On April 5, 1959, the 4907, then one of the last remaining 5 stripe GG-1's had a hole in its rear pan so it was running front pan up when seen here at North Philadelphia.

(Opposite page, bottom) Another front-pan-up clocker was caught at Monmouth Junction northbound to New York with 4904 doing the honors in January 1968. By this time, converted "City" series Pullman cars had joined the P-70 fbr's in clocker service.

(Above- Bill Volkmer, opposite page, top- Bill Volkmer, opposite page bottom- Bill Brennan)

North Philadelphia

Train Watcher's Heaven

The Northeast abounds with passenger train buffs. There was nowhere in the country that could compare with North Philadelphia station for "car counting," especially over the Thanksgiving weekend when everything that still had wheels rolled through. Aside from possibly New York's Penn Station which was underground, North Philly easily qualified as the nation's busiest in terms of passenger train throughput. Shown here is but a sampling of the action to be taken in on a typical day. *Pennsy Diesel Years* has already chronicled the diesel action here, both passenger and freight. Here is the electric:

(Above) GG-1 4876, famous as being the GG-1 that crashed into Washington Union Station in 1952 heads a southbound mail and express on October 5, 1958. Judging from the New Haven head end equipment, the train probably originated in Boston.

(Opposite page, top) A local Camden to South Philadelphia freight is hauled past the station by a lone P-5a 4728 on October 3, 1959.

(Opposite page, bottom) Pot Yard-bound 4856-4807 give waiting passengers something to stare at on the afternoon of June 23, 1959.

(All- Bill Volkmer)

NORTH PHILADELPHIA

Race Street

Engine terminal

Immediately north and west of Philadelphia's 30th Street Station, completed in 1933, lay the Race Street engine pits and its associated Penn Coach Yard, both home base for the clocker passenger fleet. In the 1960's this consisted of several GG-1's out of the pool and approximately 100 of the 162 P-70fbr coaches which had been extensively modernized in the early 1950's with new trucks, lighting, and mechanical air-conditioning. The author was given his first PRR Supervisory assignment as Gang Foreman in charge of steam heat, air brakes and air conditioning in this coach yard in 1961. Nearby was the Powellton Avenue MU storage yard where the MP-54's spent their daytime layover hours while all those famous Philadelphia lawyers plied their trades. One track in the coach yard was home base for five of the six Pennsy business cars then on the roster. The sixth car, the 7510-*Pittsburgher*, was based in the *Steel City*.

(Above) Dave Ingles shot this overall view of the coach yard and engine terminal in the summer of 1964. The engine terminal is on the right, coach yard center, repair pits to the left underneath the High Line structure. The Philadelphia Art Museum stands guard in the distance.

(Below) A view of several MP-54's being washed on July 17, 1958.

(Above- Dave Ingles, below- Bill Volkmer)

To the west of Race Street pits and Penn Coach Yard lay the ominous "High Line," a spidery steel trestle which carried the freight trains high above the passenger action and completely out of interference range of the varnish schedules. In the photo *(above)* the West Philadelphia steam plant rises as a backdrop to a pair of G's on a southbound freight headed by a very unwashed 4815 in June of 1969. Standby steam and compressed air for the coach yard and station complex emanated from this steam plant.

(R. S. Short)

30th Street Station

In the early 1930's as a part of the massive Pennsy electrification project, the railroad undertook the almost equally massive task of completely rebuilding its West Philadelphia station, coach yard, and steam roundhouse facilities. The project called for elimination of the roundhouse and replacement by the Race Street electric motor pit, building a totally new station ediface, a large MU storage yard at Powellton Avenue, a large steam generating plant for heating several PRR owned buildings including the new 32nd Street General Office building, the station, the upholstery shop and various small buildings. This led to the eventual total elimination of the Broad Street Station wherein all long distance trains either passed through or terminated at the lower level of the new 30th Street Station. The double ended characteristics of the multiple unit electric cars contributed greatly to the obsolescence of Broad Street Station and total reliance upon the recently completed Suburban (underground) Station for the commuter fleet.

(Above) The 758 leads a three car MP-54 train inbound from the upper level of 30th Street into the cavernous depths of Suburban Station in April 1969, fully a year after the PC merger.

(Below) Photographer Short stands approximately at the point where the throat to Broad Street Station once diverged. The two car train of *Pioneers* (later dubbed *Silverliner I*) is outbound towards 30th Street Station in the left background, about to cross the Schuylkill River. The two cars were part of the six original 1958 order for stainless steel MU cars and were caught here in May 1960.
(Both- Rich Short)

(Above) About the only time a locomotive-hauled train was to be seen on the upper level of 30th Street Station was for a special train or occasion. That is, after Broad Street Station was closed and torn down. One such occasion was the 25th birthday of that matriarch of GG-1's, the 4800. The National Railway Historical Society chartered a train hauled by this loco on May 17, 1959 and it began and ended its excursion at 30th Street Station upper level.

(Below) The distinctive clang, clang of the 4934's heavy-duty bell is about the only noise to be heard as a southbound train glides to a stop in the station's lower level in July 1969.
(Above- Ken Douglas, below- R.S. Short)

South Philly

Leaving the mainline at Arsenal Tower in mid-town Philadelphia, the 3.9 mile long Delaware Extension crossed the Schuylkill River on a rather spectacular bridge and trestle locally known as the Grays Ferry Bridge and paralleled the "Sure-Kill" Expressway most of its way to South Philly ore and coal piers.

Grain trains were unloaded at the Girard Point grain elevator for export and general merchandise was distributed along the Delaware Avenue waterfront to the grocery auctioneers. (*Above*) In 1948 a GG-1 in the classic pin-stripe scheme leads a freight over the Arsenal Bridge while a Reading fantrip goes underneath on the B&O tracks.

(*Nelson Bowers*)

(Above) At the Broad Street underpass we see 4760-4755. In August of 1962, the 4755 was totalled in a rear-end collision near Parkesburg, Pa. In the 4755 the day it was wrecked, the ICC Monthly Inspection certificate Form M.P. 162-E signed by the author, attested to the fact that the brakes were in proper working order. The form is reproduced on the opposite page. *(Below)* In happier days, the 4760-4755 pair emerges from under Broad Street near the Philadelphia Navy Yard, that same November 21, 1961. *(Both- Bill Volkmer)*

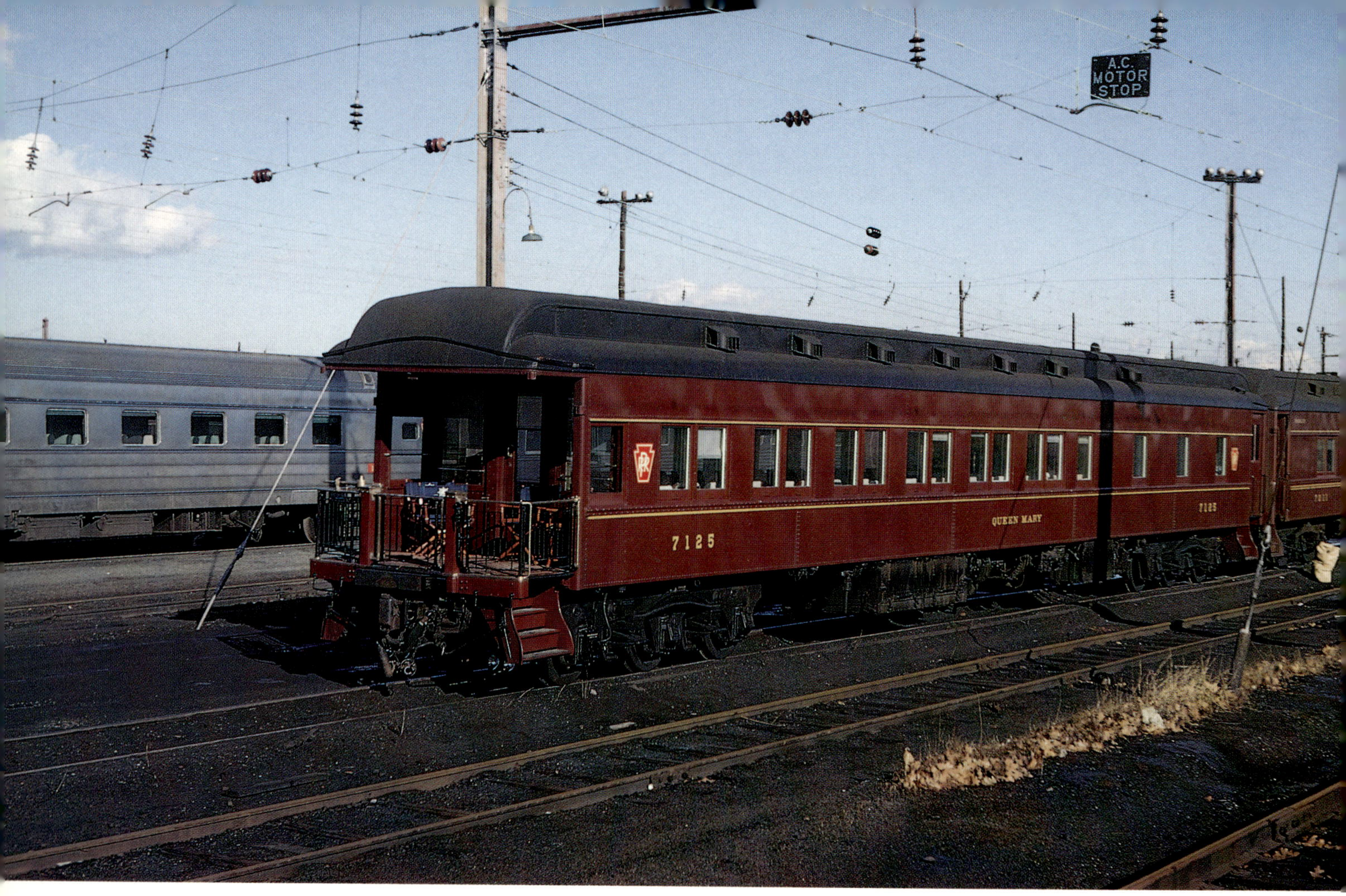
A.C.
MOTOR
STOP
7125
QUEEN MARY
7125

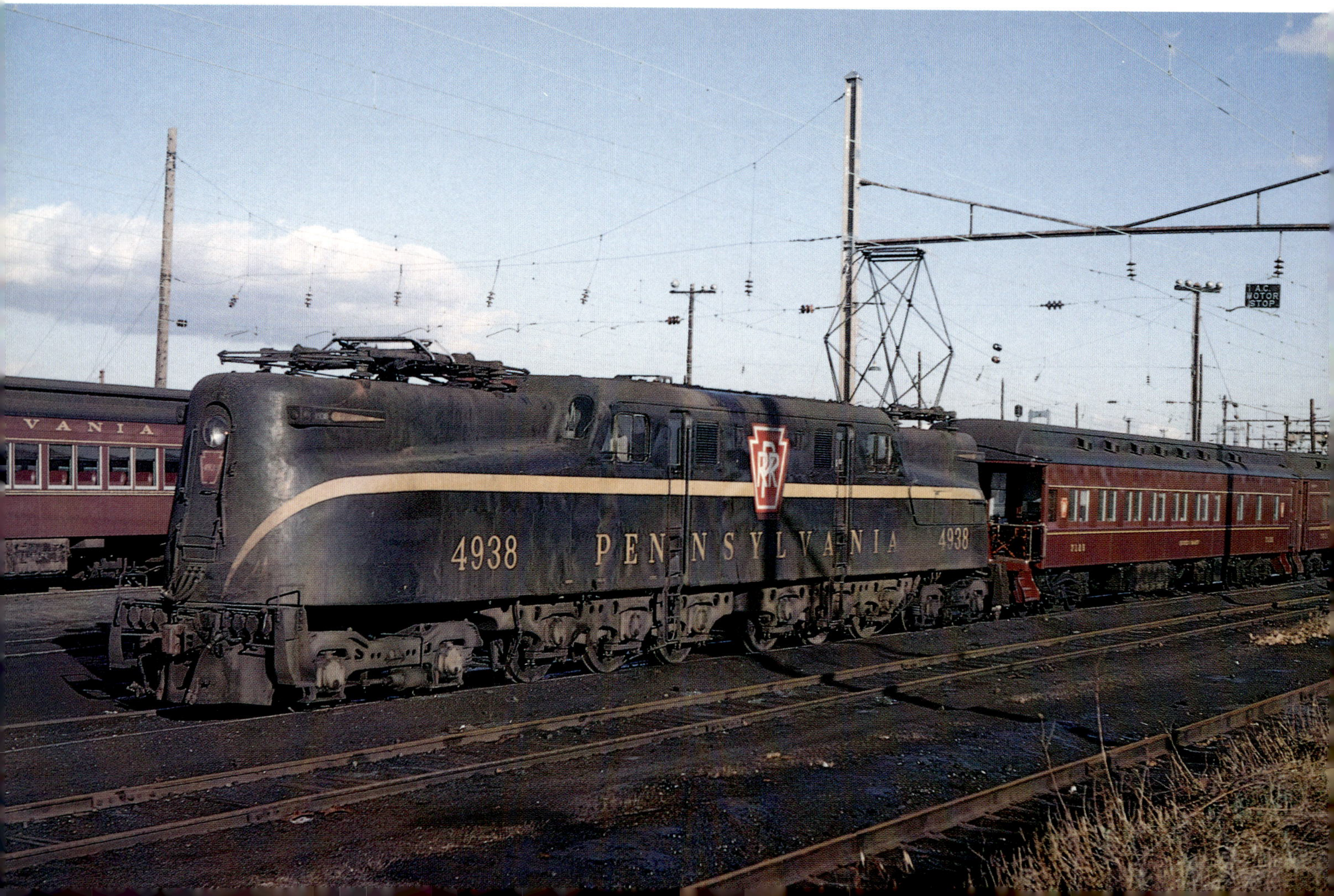
4938
PENNSYLVANIA
4938
A.C.
MOTOR
STOP

Army-Navy Game

Three hundred sixty-four days a year the South Philadelphia (Greenwich) Yard played host to freight power hauling coal, ore, grain and general merchandise to and fro. But one day per year, usually the Saturday after Thanksgiving, the chief commodity there was people. Lots of 'em.

The stadium being adjacent to the yard lent itself to a temporary Grand Central Station of sorts for the thousands who attended the annual Army-Navy football classic. It gave the Pennsy officials a chance to put their best foot forward and get paid for advertising their wares to a captive audience!

The Pennsy went all out for the occasion, making plans and holding planning sessions for months in advance to make sure everything came off without a hitch. Managers were occasionally "busted" when things went awry. There were no acceptable excuses for a late train to the game, absolutely none. Wreck crews were on standby and even called upon to extract ladies from locked restrooms where the doorknob fell out!

(Above) A view taken from the tower car platform on standby at the stadium during the game, November 26, 1960. This scene has been drastically altered over the years with the addition of an aerial expressway and removal of all catenary. Notice the polished GG-1's.

(Opposite page, top) The only surviving open end parlor-observation car in the latter years of the Pennsy was car 7125, the *Queen Mary* which was officially the Board of Directors car. About the only time the car saw service in the early 1960's was to the game. It is shown at the stadium on November 27, 1965.

(Opposite page, bottom) The 4938 is coupled onto the *Queen Mary* for the return trip after the game with the obsy on the head of the train.

(Above- Bill Volkmer, opposite page, both- Ken Douglas, Volkmer collection)

West Chester Branch

The West Chester Branch from ARSENAL Tower south of Philadelphia westward (and upward) to Media - West Chester was a little remarked commuter railroad serving as the gateway to the Octoraro Branch, the Wawa Branch, the Chester Creek Branch and the Frazier Branch. Service on the line, typical of commuter lines elsewhere, was heavy in the rush hours and either single *Silverliner* or two unit MP-54 cars on half hour headways at other times.

(Left) Car 153, one of the original *Pioneer* cars built by Budd in June 1958 pauses at the Clifton-Aldan Station where the *Red Arrow* Sharon Hill line trolleys connected. The date, September 1, 1962.

(Below) Media, in pre-SEPTA days was the terminus of all Philadelphia trains. Connecting cars for West Chester layed over on the track at the right. On October 5, 1958 there was service to West Chester only in the morning and evening except for a single mid-day round trip.
(Both- Bill Volkmer)

The territory west of Media (southward by railroad) served by the electrified West Chester Branch was a sparsely settled, highly rural scene, a far cry from today's densely populated suburbia, typical of other commuter runs around the nation. Probably the only reason the line was electrified beyond Media was to avoid an engine change at Media. Indeed, nearly all through passengers were required to change trains at Media as the Pennsy seldom ran through service from West Chester to Philadelphia. Travel time from Suburban Station to West Chester was 65-70 minutes including a change of trains.

(Above) A single outbound MP-54 makes the flag stop (as were all stops on the West Chester line) at Westtown, Pa., the next to the last station. The mid-day run was photographed June 22, 1957.

(Below) A lone MP-54 E6 sits forlornly at the now-seedy West Chester station August 25, 1965 awaiting the afternoon departure for Media. This was surely one of Pennsy's most obscure MU operations.
(Above, Nelson Bowers, below, Tony Schill)

Wilmington Shop

The "temporary" shop
that became Permanent

In the early years of Pennsy electrification, it was always a goal of management to reach Pittsburgh with the overhead catenary. If and when that goal was reached, it would only be natural that Altoona would be the chief maintenance point for the locomotives. After all, that's where they were all built at the time. As an interim solution, an existing steam locomotive backshop at Wilmington, Delaware was converted for the New York to Washington fleet (i.e. the GG-1's) and Enola roundhouse became home base for the freight P-5a's. Because GG-1's were cumbersome on a turntable and lent themselves to an overhead crane, Wilmington Shop was ideal. It was not ideal, however, from the standpoint that it was single ended; the kiss of death in shop design. To boot, Wilmington from an operations standpoint, was inaccessible from all directions. Passenger trains only paused there momentarily, none originated there, and a very low percentage of freights had cause to enter Edgemoor Yard. So because of the advent of diesels and scrapping of the plans to electrify to Pittsburgh, Wilmington Shops became the permanent home of the GG-1's, and the various experimental electric locomotives in the ensuing years.

(Above) One such experimental was the 1939 GG-1 freight clone, the DD-2 #5800. It was the prototype freight version of the GG-1 and had the electrification been extended to Altoona, it may or may not have been the locomotive of choice. The 5800 spent its last days alternating between the Baltimore Tunnel Helper and the once-a-day local from Edgemoor to South Philadelphia. This photo shows 5800 in back of the Wilmington Shop on April 4, 1962.
(Fred Cheney)

Edgemoor Yard

Adjacent to the Wilmington backshop was the Wilmington Enginehouse and Edgemoor Yard. Edgemoor served as a classification point for freight destined to and from the Delmarva Peninsula. Electric power was used on trains dispatched from here to Camden, South Philadelphia, Enola, Baltimore and Potomac Yard. Many freights changed power here for the purpose of getting it into and out of the Wilmington Shops for maintenance.

(Above) The two unit Westinghouse rectifier set, 4995-4997 is departing northbound from Edgemoor and will make a call at South Philadelphia to drop the horse manure seen four cars back. That product of nearby racetracks was destined for the mushroom farms near West Chester only 30 miles distant by highway but closer to 100 by rail, not to mention numerous interchanges between local freights. Small wonder the trucks have all that business today. Photo taken August 19, 1961.

(Below) E-2b's 4942-4943 have just arrived with a mixed freight, March 21, 1959. All maintenance work on the rectifiers and G.E.'s was performed at Wilmington Shops.
(Both- Bill Volkmer)

The Shellpot Branch

There were running tracks which served Edgemoor Yard and, in effect, kept the freight trains separate from passenger trains around the Wilmington passenger station, at least those needing to make setoffs and pickups at Edgemoor. These tracks were known as the Shellpot Branch and the Shellpot Secondary Tracks. About midway along the seven mile line was a trestle and bridge over the Christiana River near the point where the Reading Railroad crossed the Pennsy. A block station, often manned by our own Rich Short, called TRESTLE by the Reading and BRIDGE by the PRR, controlled movements across the Reading (which ran down to a carfloat across the Delaware River) and onto the Newcastle Secondary for Delmarva-bound trains. The block operators were Reading employees working to Pennsy rules, and some of these block operators could often be found taking pictures of passing Pennsy freights! Herewith, we show two:

(Above) The 4454 leads a northbound freight into Edgemoor Yard in May 1965 while *(below)* a pair of G's are seen passing the block shanty. One had to be careful not to step out the door too quickly when one of those silent GG-1-powered trains was approaching! This photo taken in June 1965.
(Both- Rich Short)

Ivy City

Washington, D.C., was the southernmost passenger destination on the electrified Pennsy, but only a fraction of the total patronage in those days boarded or left the trains there. Rather, for most, DC was merely an engine and crew change point where coaches were added, subtracted or transferred to southbound trains operating under flags reading SR, C&O, SAL, and ACL. Ivy City Engine Terminal, about three miles north of Union Station, was a joint facility of the Washington Terminal Co., the B&O, and PRR. Pennsy's GG-1's could be seen laying over there in the company of E-6's, E-7's and E-8's and numerous passenger F's of the various other roads. Altogether a really interesting place. The 4901 *(above)* hustles the northbound cars of the C&O's FAST FLYING VIRGINIAN on a bright April 9, 1961 while C&O and Southern diesels are serviced for hauling those same coaches south in the evening. *(Below)* The 4899 was herself being serviced near the Ivy City turntable October 18, 1958. *(Both- Bill Volkmer)*

Second Hand Rose

In this era of mega-mergers, instant shortlines, rent-a-locomotive, and locomotives being bought, sold, and traded like so many cattle at a livestock auction, it is difficult to imagine or remember the fact that the Pennsylvania Railroad almost never bought a second-hand steam, diesel, or electric locomotive and equally seldom, ever sold a locomotive of any kind for continued railroad use. (Three second hand diesel purchases were recorded: F3B's from BAR, a Baldwin from P&WV, and RS2's from D&H).

So, the aquisition of the second-hand Great Northern electric locomotives in 1957, previously described in this book, was truly a rare occasion in Pennsy history. That these locos were shortlived in PRR livery was also an unusual turn of events.

(Above) Shown here is Great Northern 5014 on its arrival at Altoona Works in 1957 before conversion to PRR standards (i.e. cab signal equipment added and removal of side doors and the bus jumpers from the roof ends). It is interesting to note that no PRR electric locomotive or MU car was ever electrically connected to its running mate by 11,000 volts as, for example, the Reading did. *(Below)* GN #5014 became PRR #4 which rests at Juniata Shop awaiting motor work in September 1959. The motors were never rewound as rewind kits would doubtlessly have had to be custom fabricated, and the loco was cut up for scrap on June 27, 1960. Salvaged parts were then applied to sister unit # 2 so it could run a very few more miles. All FF-2's were in permanent dead storage at Enola prior to April 1962.
(Above- Bob Watson, below- Dave Sweetland)

Bride and Grooms

Most of the Pennsy MP-54 fleet was built in two large groups. The original Paoli fleet was converted from then fairly new coaches at Altoona Works in 1912-1913. PRR had designed the P-54 coach to be its standard steel coach (54 feet long) in 1908 and had structurally prepared them for eventual carriage of a transformer underneath and a pantograph mounted on the roof. They even included the portholes in the end sheets for the motorman's windshield. The second large group was converted in the 1926 time frame and at that time 40 new coaches were built from scratch. These were known as MP-54 E-1 and E-2 respectively. Between 1932 and 1937, while the GG-1's were being built, a group of 43 motor-trailer combinations were rebuilt from steam coaches and turned out as semi-permanently coupled pairs. These were officially classed as MP-54 E-3 but unofficially they were dubbed the "bride and grooms." The trailer unit was nearly always a coach configuration with a few RPO's thrown in for good measure. The motor units had traction motors utilizing armatures interchangeable with the GG-1 motors! The motored units of the bride and grooms consisted of combines and straight coaches.

(Above) Trailer 34 leads a three car train onto the Chestnut Hill Branch at North Philly on July 22, 1961. *(Below)* Combine 4574 lays over at Trenton. Sister combine 4561 was notable for being the first railway car in history to be outfitted with an ignitron rectifier in July 1949.

(Above- Bill Volkmer, below- Marty Zak, Yanosey collection)

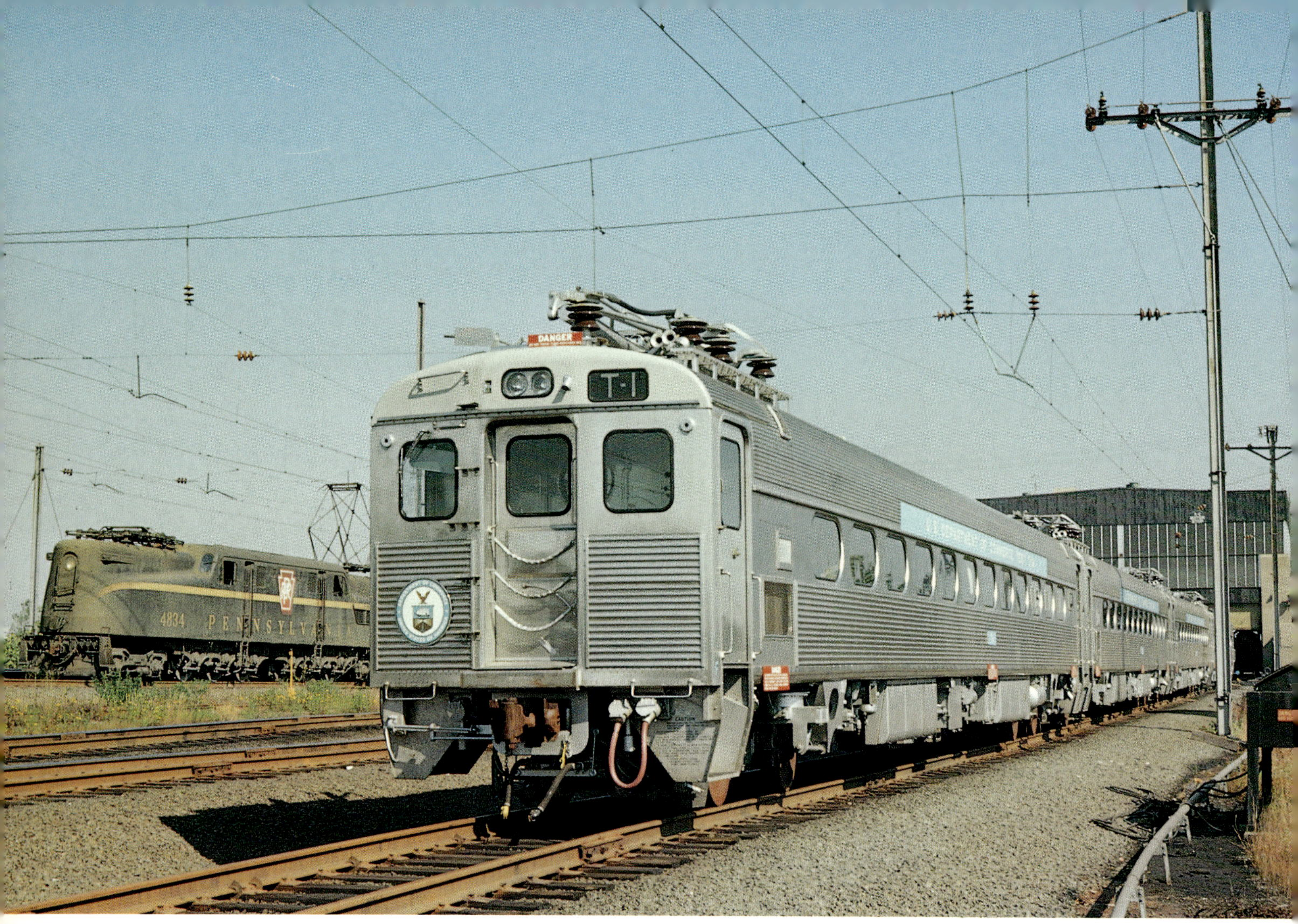
DANGER
T-1
4834
PENNSYLV

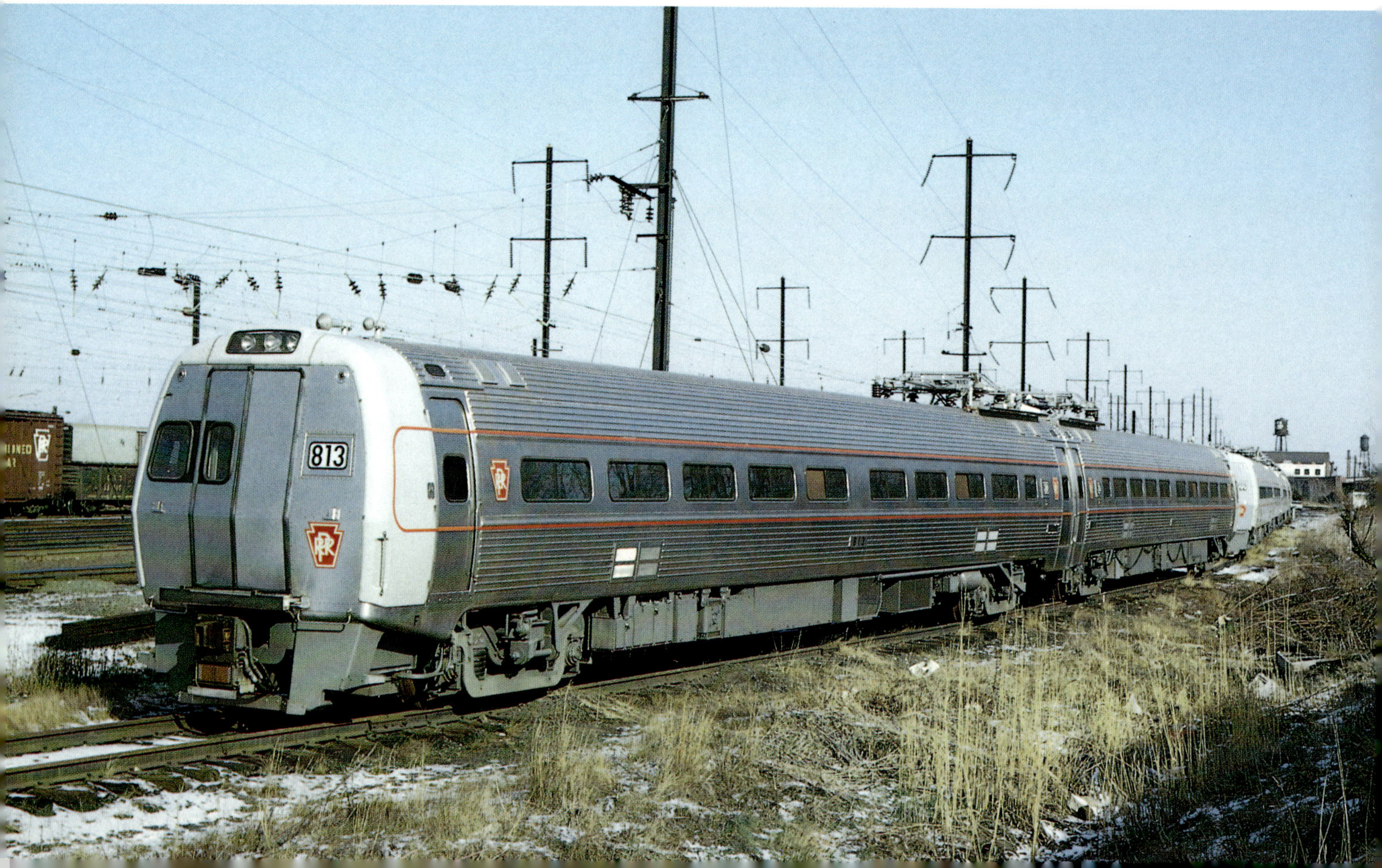
813

The Battle of
the locomotive concept
versus MU trains

By the late 1950's, the Pennsy had dieselized the non-electrified portions of the railroad and the long distance passenger train appeared to be doomed with the 1957 advent of jet powered commerical aircraft. At *6 Penn Center Plaza,* a decision had not been finalized on whether to leave the wires up on the freight lines or pull them down in favor of diesels. Pennsy had been wrestling with this decision for over a decade while the diesels were being procured for the western lines. During this time, the New York-Washington Corridor coach fleet and the GG-1's came to be in dire need of replacement. No new coaches and parlor cars had been purchased since the 1952 overhaul of the CONGRESSIONAL LIMITED except for the 1956 aquisition of the Budd tubular train.

There was a severe shortage of cash available for capital improvements due to a sluggish economy, purchase of 2,465 diesels, and mounting revenue losses in providing the passenger services that remained. Therefore, year after year, there was no money made available for either new coaches or locomotives equipped for passenger service. Since the concept of electric-powered MU cars appeared here to stay, insofar as commuter train service was concerned, it made a certain amount of sense to convert the New York-Washington corridor to an all MU passenger operation utilizing high speed equipment over vastly improved track and roadbed structure.

After the 1959 decision to buy E-44 freight locomotives and the 1963 Philadelphia government aquistion of MU cars, it became evident that the electrification abandonment plans were off, so Pennsy was successful in lobbying Congress to appropriate funds to upgrade the Northeast Corridor in the public interest. Congress passed the Transportation Improvement Act in 1965 funding the project with car aquisition first and roadbed improvements later.

The big problem was that the arch-conservative Pennsy management forsaw at least a ten year test program before any reliable high-speed MU cars could be made available. Memories still lingered of the O-1 / P-5 debacle of the early 1930's where too little testing had been performed before a large quantity of locos was ordered. The public and the Federal Government would not stand for 10 years of testing and insisted that only one or two years at most be devoted to developing a suitable car.

In 1966 the Budd Company built four test cars numbered T-1 to T-4 and these were tested first on the Reading out of Jenkintown and later on the Pennsy out of Morrisville. In 1966, an order for 50 *Metroliners* was placed with the Budd Company and a later order added 11 more cars for a Harrisburg service with MU car equipment.

The cars were delivered in 1967 with red Keystones affixed. Alas, the railroad paid a price when trying to compress 10 years of testing into 3 and the *Metroliners* had not entered revenue service by merger day, February 1, 1968. Indeed it was into the spring of 1969 before the *Metroliners* started hauling revenue passengers.

(Opposite page, above) Test car T-1 sits at Morrisville enginehouse on October 9, 1966. Bob Watson, incidentally, whose photography was seen elsewhere in this book, was the principal test engineer representing the Pennsy interfacing with the USDOT engineers who were actually engaged in the testing activities. *(Opposite page, below)* Metroliner #813 idles the time away at Trenton between test runs and retrofits on January 11, 1969. By this time the PC merger was almost a yead old, but there was little incentive to replace the Keystones with PC "worms" because the car had not yet entered revenue service. A few months later the cars entered revenue service, but the bugs remained to some degree for many months to come.

(Both- Allan Roberts)

D.C. and Potomac Yard

South of Washington Union Station the Pennsy freight electrics threaded their way among the bureaucracies of the nation's Capital giving the bored government office workers something to stare at out the window. Destination of the southbound freights lay just across the Potomac River in Alexandria, Virginia: Potomac Yard. Pot Yard, as it was known, was the principal freight interchange point between the Pennsy and all of the various southern railroads via the bridge line railroad, Richmond, Fredericksburg and Potomac. Freights originating at Pot Yard fanned out to all the major PRR terminals of the electrified territory but the preponderance of the traffic was destined for the Jersey City area and points north.

(Below) On May 19, 1967 a pair of E-44's led by the 4458, by now converted to class E-44a through addition of solid state rectifier equipment hauls a Pot Yard-bound freight past the government office buildings in downtown Washington.

(George Menge, Volkmer collection)

Maintenance

Thus far in these pages, we have attempted to provide a picture of what the electrified portion of the PRR looked like, give a little of the history of how the locomotives evolved, and with the aid of photographs, provide some insight as to how and where the railroad operated.

On the next few pages, we'd like to show a few glimpses of the "behind the scenes" maintenance activities which were carried out day after day in order for the motors to keep humming and the wheels turning. It was as a result of the dedication of the maintenance forces, working in hot summer heat, blowing snow, sleet, pouring rain, and bone-jarring cold weather that the railroad kept running. The author saw this all first hand and can offer testimony to this dedication. The maintenance men seldom had first class working conditions. Indeed, the financially strapped railroad nearly always was forced to render the maintenance forces as second class citizens during the budget cutting process.

The electrified territory often presented more of a challenge to maintenance than its non-electrified counterpart. If the catenary system failed for any reason, the trains usually ground to a halt. It was that simple. For example, on three occasions during the winter that the author worked at Enola electric pit, the trolley wire burned through during sleet storms. This was caused by heavy wet snow weighting the pantograph shoe away from the ice-coated wire and causing an arc to draw between the wire and the shoe. As a result, the wire burned in two. The process of bleeding the air on twenty or so P-5's and GG-1's and towing them out from under the broken wire so they could be inspected and dispatched, using non-electrified tracks and diesel switchers, was about a six hour ordeal, not to mention the disruption caused by the wire train putting the wire back up!

(Above) P-5a 4719 and E-2C 4998 are seen on the inspection pit at Kearny Meadows, N.J., in March 1962. To the right of the photo was a building housing a sand drying stove which required the services of a man round-the-clock who shoveled wet sand into it. The stove-dried sand fell out the bottom into a pit below and was then blown by compressed air into the overhead towers seen in the photo. *(Bill Brennan)*

Life under the Wires

One of the big reasons Pennsy and other railroads were soured on electrification was the tremendous infrastructure of catenary, transmission lines, substations, power generating stations, and indeed even the maintenance trains themselves which all were in constant need of maintenance. The author, as a part of his training in PRR management spent some time with the Traction Power Department crews learning first hand how the physical plant was maintained. Herewith is a glimpse behind the scenes of maintenance, electrification style.

Most labor intensive was the catenary, inspection of which was carried out somewhere on the system every working day of the year. Inch-by-inch the folks *(left)* on the tower car would check each hanger clip, each insulator (a prime target for gunners), and each foot of copper for burn-throughs caused by arcing pantographs in sleet. April 8, 1960.

(Opposite page, top) Tower car 489502 at Daylesford, just east of Paoli on April 8, 1960.

(Opposite page, bottom) The same car from the opposite end. The raised pantograph did not collect power. Rather, it provided a direct ground connection to the running rails to insure that no juice was flowing in the wire being inspected.

(All- Bill Volkmer)

Catenary Nomenclature

Few railfans bother to understand the workings of the catenary system, other than the fact that the trolley wire is the one the pantograph rubbed on and transmitted power to the locomotives. Actually, as illustrated in these pages, it took a number of wires, each doing its intended job, to deliver that power.

The uppermost pair of wires shown in these photos suspended by giant vertical insulators were the 132,000 volt A.C. "high voltage" transmission lines. This power was stepped down periodically at small substations to the 11,000 volt A.C. needed to power the locos and MU trains.

The horizontal wires between the opposing catenary poles were called "cross spans," the gracefully sagging wire directly above the trolley wire was the "messenger wire," and the vertical wires, "hangers." Horizontal insulators were the "steady" insulators and vertical insulators, "suspension" insulators.

Much of the celestial knitting which proliferated the landscape had nothing to do with powering the trains. Rather, these were wires used for controlling signals, wires for the railroad's own telephone system, signal transmission lines, and in the early years, telegraph lines.

(Below) GLEN Interlocking at Glen Lock, Pa., had enough tracks and signals present to exemplify the intricate catenary network as this location was the point of departure of the Trenton Cut-Off from the mainline, five and one half miles west of Paoli. Here, a lone unidentified GG-1 lays sand on its ascent of the grade from Thorndale to Paoli, the top of that grade, in March 1956.

(Opposite page, top) On a typical hot, muggy, August afternoon in 1966 the camera of Tony Schill catches people waiting at Elizabeth for the approaching MU train while a fleeting southbound GG-1 4903-powered train throws them a blast of hot, but dry, air to somewhat ease the uncomfortableness of the day. This photo illustrates the process of hanging catenary on a curve, wherein the messenger wire does double-duty.

(Opposite page, bottom) No curve here. The straight-as-an-arrow track through Lawrenceville, New Jersey allows the 4923 to effortlessly hold the speedometer needle on the 80 mark as she hurls the SILVER METEOR towards New York in December 1966. The horizontal bar about halfway up the catenary pole carried the signal transmission lines.
(Below- Nelson Bowers, opposite page, top- Tony Schill, opposite page, bottom- R.S. Short)

Wrecks

Another reason the Pennsy mechanical forces disdained the electrification was the difficulty with which wrecks were cleared.

(Above) Witness the low booms on the derricks under the catenary at this derailment on the Trenton Cut-Off near Malvern, Pa., on April 15, 1960. Power had to be shut off on all tracks while the derailed cars were carefully lifted. Anyone knowledgeable in physics will agree that a low boom is bad news.

(Below) A one-of-a-kind derrick was this double ended, electric (third rail) powered monster which reposed at PRR's Sunnyside Yard when not retracking trains in the New York tunnels where the clearances were tight. The derrick was numbered 490797 and the photo was taken March 11, 1960. *(Both- Bill Volkmer)*

(Above) The 489502, a former doodlebug was depicted after the car had been wrecked by colliding with an REX car in 30th Street Post Office (brake failure). A year later on May 25, 1961, the car emerged from Wilmington Shop resplendent with a new coat of paint and a new front end. Shown here at 46th Street Enginehouse in Philadelphia. During its year of absence, wire inspection costs skyrocketed because a full crew was required to man the wire train for inspections.

(Below) The 6 spot has spun itself out of its tires (see FF-2 page) and dropped to the crossties. Location is Thorndale and the date, August 1960. *(Above- Bill Volkmer, below- Les Broomfield, R.S. Short collection)*

Museum Bound

At the end of the roster section of this book we have attempted to list all of the known PRR electric locomotives which are currently preserved in museums. The 4700, 4800, 3936-37, 4876, and 4903 are depicted in this book as they appeared during their Pennsy days. There is a strong possibility that one or more E-44's will ultimately be selected for preservation as well. On these pages, we'd like to pay special homage to two of the more notable specimens which have been preserved. *(Above)* The 4780-81 are seen near "Q" Tower at Sunnyside Yard in March 1968, just prior to retirement. After their interment at the Railroad Museum of Pennsylvania *(left)* they were restored to their original 3936-37 numbers where they were photographed on June 20, 1987. *(Above- Allan Roberts, below-Dave Burnette, Volkmer collection)*

(Above) In December 1977 the National Railway Historical Society, led by their Philadelphia Chapter, raised funds to pay Amtrak the costs incurred in returning GG-1 4935 back to its original PRR five-stripe paint scheme appearance. Consequently the loco was brought into Wilmington Shop, very soon after this photograph was taken at Wilmington inspection pit on January 22, 1977.

(Below) Three and one-half months later on May 7, 1977, it emerged from that shop and when the author took this photo, the painter was still on the opposite side touching up the lettering! The 4935 finished out its last four years of Amtrak service appearing just as it had for its first twelve years. When it was finally retired it was sent to the Railroad Museum of Pennsylvania in Strasburg where it resides today. *(Both- Bill Volkmer)*

PRR ELECTRIC LOCOMOTIVE ROSTER

Numbers	Class	Wheel Arrangement	Builder	Date Built	Date Scrapped	Notes
1-7	FF-2	2-6-6-2	G.E.	1926-30	1960-63	1
3900-3901	B-1	0-6-0	PRR	1926	1962	
3910-3921	B-1	0-6-0	PRR	1926	1958-62	2
3922-3929 *	L-5a	2-8-2	PRR	1924-26	1942	
3930	L-5	2-8-2	PRR	1924	1944	
3931	FF-1	2-6-6-2	PRR	1917	1940	3
3932-3949 *	DD-1	4-4-4-4	PRR	1910-11	1940-62	4
3950-3951 *	AA-1	0-4-4-0	PRR	1907	sold-LI	5
3952-3995 *	DD-1	4-4-4-4	PRR	1910-11	1940	4
3996-3999 *	DDodd	4-4-4-4	PRR	1909-10	1940	
4400-4459	E-44	C-C	G.E.	1960-63	to PC	
4460-4465	E-44a	C-C	G.E.	1962-63	to PC	
4700	P-5	4-6-4	PRR	1931	preserved	6
4701	P-5a	4-6-4	B-W	1932	1960	
4702	P-5b	4-6-4	B-W	1932	1961	7
4703-4742	P-5a	4-6-4	B-W	1932	1962-65	
4743-4754	P-5a	4-6-4	B-W	1934-35	1961-62	8
4755-4769	P-5a	4-6-4	G.E.	1932	1961-65	
4770	P-5a	4-6-4	G.E.	1932	1962	9
4771-4774	P-5a	4-6-4	G.E.	1933	1959-63	
4775-4790	P-5a	4-6-4	PRR	1935	1961-62	8
4791	P-5	4-6-4	PRR	1931	1950	10
4800	GG-1	4-6-6-4	G.E.	1934	preserved	11
4801-4814	GG-1	4-6-6-4	G.E.	1935	to PC	
4815-4938	GG-1	4-6-6-4	PRR	1935-43	to PC	
4939-4942	E-2b	B-B	G.E.	1951	1964	
4943-4944	E-2b	B-B	G.E.	1951	1964	12
4995-4996	E-3b	B-B-B	West.	1951	1964	
4997-4998	E-2c	C-C	West.	1951	1964	
4999	R-1	4-8-4	B-W	1934	1956	13
5684-5697	B-1	0-6-0	PRR	1934	1960 part.	
5800	DD-2	4-4-4-4	PRR	1938	1962	
5938-5939	L-6	2-8-2	PRR	1932	1967	14
5940	L-6a	2-8-2	Lima	1933	1967	15
7801-7807 *	L-5pdb	2-8-2	PRR	1927	1942	
7808-7811 *	L-5pdg	2-8-2	PRR	1927	1942	
7812-7815 *	L-5pdw	2-8-2	PRR	1927	1942	
7850-7851	O-1	4-4-4	PRR	1930	1948	
7852-7853	O-1a	4-4-4	PRR	1930	1949,61	
7854-7855	O-1b	4-4-4	PRR	1930	1948	
7856-7857	O-1c	4-4-4	PRR	1930	1949-61	
10003	2B	0-4-0	B-W	1907		16

* Denotes loco D.C. operated.

Notes

1- Ex-Great Northern 5011-5017 purchased 1957. 5018 purchased for parts. EMD noses used in 5854 E-7 and 9859 FP-7.
2- Following units renumbered in 1956: 3910/4750,3912-13/4751-4752,3919/4753,5685/4754,5687/4755,5690/4756,5693/4757. 4750 preserved.
3- Nicknamed "Big Liz." Used as a Paoli helper. Second PRR attempt to build an A.C. freight loco.
4- 46 units sold to Long Island R.R. in 1927-1944.
5- 3950 Sold to Long Island 323 in 1916- Originally numbered 10001-02. D.C. locos originally tested on the WJ&S.
6- Originally numbered 7898. Renumbered 4700 in 1933.
7- Originally P-5a. Rebuilt in 1937 with motorized pony trucks
8- Modified carbody style.
9- Originally a boxcab. Rebuilt to modified in January 1945.
10- Originally numbered 7899. Renumbered 4791 in 1933.
11- Originally numbered 4899. Renumbered 4800 in 1935.
12- Originally built as demonstrators in October 1951 for the Great Northern R.R. Sold to PRR in March 1953.
13- Originally numbered 4800. Renumbered 4899 in 1935. Renumbered 4999 in 1940.
14- Originally numbered 7825-7826. Renumbered 5938-5939 in 1933. 5938 scrapped 1966, 5939 renumbered 4790 in 1966, scrapped in 1967.
15- Renumbered 4791 in 1966, scrapped in 1967. 29 additional carbodies constructed and stored several years at Altoona until scrapped in 1942.
16- First PRR A.C. loco. Test operated on Long Island R.R.

List of Preserved Electric Locomotives as of 1-1-91

```
4465     E-44    R.R. Museum of Pennsylvania, Strasburg,Pa.
4700     P-5     National Museum of Transport, St.Louis, Mo.
4756     B-1     (Orig.5690) R.R. Museum of Penna.,Strasburg,Pa.
4780-81  DD-1    (Orig. 3936-37) R.R. Museum of Penna.
4800     GG-1    (Orig.4899) R.R. Museum of Penna.,Strasburg,Pa.
4859     GG-1    (First electric loco to Harrisburg) Hbg. Station.
4872     GG-1    New Jersey Transit.
4873     GG-1    Private Owner, Philadelphia, Pa.
4876     GG-1    New Jersey Transit.
4877     GG-1    New Jersey Transit.
4879     GG-1    New Jersey Transit.
4882     GG-1    Lake Shore R.R. Museum, Elkhart, Ind.
4890     GG-1    B&O R.R. Museum, Baltimore, Md.
4903     GG-1    National Museum of Transport, St. Louis,Mo.
4913     GG-1    Railroader's Memorial Museum, Altoona,Pa.
4917     GG-1    Leatherstocking Ry. Museum, Cooperstown Jct.,N.Y.
4918     GG-1    Smithsonian Institute, Washington, D.C.
4927     GG-1    Illinois Ry. Museum, Union, Ill.
4933     GG-1    Central N.Y. Chapter, NRHS, Marcellus, N.Y.
4935     GG-1    R.R. Museum of Pennsylvania, Strasburg,Pa.
```

The BROADWAY

in twilight

In this, our closing photo, the BROADWAY LIMITED is in its "twilight." But wait, its only 9 o'clock in the morning! That's because the train is late, hurrying past Morrisville enroute to its New York destination, still some 60 odd miles distant. By September 1967, the elegant lady, Train 28, had lost its RPO car in favor of a single B-60b mail storage car. Stainless steel coaches had been added to degrade the train's long time all-Pullman status. This was brought about by the demise of its sister train, the GENERAL. The symbolic Keystones, always synonymous with the BROADWAY's image, had less than five months to live, and indeed, the very institution of railroad-operated passenger trains would succumb (to Amtrak) three and a half years hence, in May 1971!

This photo by Rich Short, however, typifies the timelessness of the Pennsy electrification throughout its 59-plus year history. Broad sweeping curves, the gracefulness of a ubiquitous GG-1, the red Keystones, a name passenger train, all continous parts of the Pennsy electric scene. The weeds encroaching upon #1 track were also symbolic of the deferred maintenance times signifying the approaching end. We who fondly recall the Pennsy try, to erase these small glitches from our memory banks and only try to recall the good times. It is our fervent hope that this book has done the job of helping its readers recall the good times of the . . .

Pennsy Electric Years.

80+94
16-7 Jan 16
6 May 23

68-9
77
126-7

cars, an unusual low "stepless" type manufactured by J.G. Brill, were purchased in 1928 and 1929. Bus conversation began about the same time, with the major lines converted to trackless trolleys by 1940. The details of both cars and routes are presented in separate chapters.

The last third of the book covers the line to New Castle and Delaware City, a hard-luck adventure if ever there was one; the tiny, one-track Odessa & Middletown Railway; the connecting lines by which one could reach West Chester; and lines proposed but which never ran, including those on Virginia's Eastern Shore.

The book is amply illustrated, with clearly reproduced photographs. A fold-out insert map shows details of trackage for most of the routes, and there are several smaller track maps in the text. A city street map, with the lines superimposed on it, would have been helpful, but the reader can follow the routes with the maps provided. Layout is in triple columns, the text is readable and a mass of detailed information is provided.

This book breaks new ground in covering a previously neglected region of electric transit and, with its modest price, will make a welcome addition to any traction student's library.

J.N.J.H.

PENNSY ELECTRIC YEARS
by William D. Volkmer

Morning Sun Books, 11 Sussex Court, Edison, N.J. 08820. 8½" x 11", 128 pp., 1991. $45.00 (plus $3.50 postage and handling).

This book, according to its author, has two purposes: to supplement the all-color four-volume *Pennsy Diesel Years* series, and to present an insider's view of the railroad and its electrification. The geographical layout familiar to readers of the diesel series is compromised here, probably because of the compact multi-route structure of the Pennsylvania Railroad's electrified lines, but the sharp color photographs and the extensive captions remain distinguishing features of this book.

In addition to the photographs and descriptions, the book contains a short introductory text, a chronology of milestones concerning electrification from the earliest testing (1895) to the merger with New York Central (1968), a roster of electric locomotives owned by the PRR and a list of those locomotives still in existence today — in museums or privately held, including some GG1s still owned by New Jersey Transit. The time frame is largely the last decade of PRR's life, a decision which omits some of the pioneers. To be fair, however, the true pioneers died before color photography could capture them in action and most of the others had representatives from their classes still around and thus included here, if only minimally.

The book is at its best, however, not in the quality or variety of the pictures but in meeting the second goal stated above. Volkmer was very familiar with the electrified divisions of the Pennsy, and provides insights which could not be gained by any amount of library research or searching through photographic collections. In addition, many of the pictures could only have been taken by an insider. The pictures and description of catenary and its repair are particularly interesting, as are several views deep within Sunnyside Yard, New York.

As the author freely admits, little is told about the locomotives themselves which is not already published elsewhere, but the variety of color shots illustrating those facts is a strong point of this book. I have two rather minor quibbles: the map of electrified lines could use greater enlargement of the many lines around Philadelphia and New York (which the relevant text could then make reference to), and the book's organization is unclear, not being totally geographic nor chronological nor by model type. However, its strengths outweigh its weaknesses, and I recommend it to all interested in the PRR as well as those interested in railroad electrification in general.

Richard G. Prince

THE HUB OF BURLINGTON LINES WEST
by Alfred J.J. Holck

South Platte Press, P.O. Box 163, David City, Nebr. 68632. 8½" x 11", 380 pp., 1991. $59.95 (plus $3.50 postage).

Lincoln, Nebr. was to that part of the Burlington west of the Missouri River what Galesburg, Ill. was to the lines east thereof. This book concerns the development (beginning generally around 1870), growth, operations, and in many cases the decline or abandonment of Burlington's lines in the southeastern quadrant of Nebraska. While emphasis is on the trackage comprising the Lincoln Division of the CB&Q, this division's boundaries changed many times between formation of the Burlington Lines West from its component parts and the 1970 merger into Burlington Northern. The book at least mentions all trackage, including some in Iowa and Missouri, that however briefly was part of the Lincoln Division at any time.

Early chapters introduce the reader to the component parts of Lines West, chief among them being the Burlington & Missouri River. Longer chapters detail their combining and merger into the Burlington. There follow chapters on operations, power, passenger and freight trains, the seemingly-requisite chapter on disasters, details on division changes and the individual lines and personal recollections by retired railroaders. Many pictures, generally very well reproduced, illustrate virtually every facet described in the text, and many times amplify details which, if all were incorporated into the text, would make for dry reading. Well-drawn, complete, and appropriately-placed maps illustrate a variety of locations and eras. Finally, appendices show numerous public and employee timetables for the region over time, as well as locomotive assignment documents and a copy of a circular describing the "non-stop **Zephyr** run" from Chicago to Denver on October 22, 1936.

The reading is easy and, although seemingly sketchy in places, does give the overall flavor of the topic, and is regularly amplified by picture

Continued on page 39